Polynomials - Exploring Fundamental Mathematical Expressions

Edited by Mudassir Shams and Bruno Carpentieri

Published in London, United Kingdom

Polynomials - Exploring Fundamental Mathematical Expressions
http://dx.doi.org/10.5772/intechopen.1003393
Edited by Mudassir Shams and Bruno Carpentieri

Contributors
Andrei Popescu, Asan Sydygalievich Omuraliev, Bruno Carpentieri, Clark Kimberling, Claudiu Hapenciuc, Constantin Volosencu, Cristian N. Mihăilescu, Ella Dayrbekovna Abylaeva, Elyas Arsanjani Toroqi, Faiza Akram, Hashem Saberi Najafi, Laiba Khalid, Lucian Milica, Mihai Oane, Moin-Ud-Din Junjua, Mudassir Shams, Peyil Esengul Kyzy, Saima Akram, Seyed Reza Mirshafaei

First published in London, United Kingdom, 2025 by IntechOpen
IntechOpen is the global imprint of INTECHOPEN LIMITED, registered in England and Wales, registration number: 11086078, 167-169 Great Portland Street, London, W1W 5PF, United Kingdom

For EU product safety concerns: IN TECH d.o.o., Prolaz Marije Krucifikse Kozulić 3, 51000 Rijeka, Croatia, info@intechopen.com or visit our website at intechopen.com.

British Library Cataloguing-in-Publication Data
A catalogue record for this book is available from the British Library

Polynomials - Exploring Fundamental Mathematical Expressions
Edited by Mudassir Shams and Bruno Carpentieri
p. cm.
Print ISBN 978-0-85014-981-4
Online ISBN 978-0-85014-980-7
eBook (PDF) ISBN 978-0-85014-982-1

If disposing of this product, please recycle the paper responsibly.

Meet the editors

Mudassir Shams received his Ph.D. in Computational Mathematics (Numerical Analysis) from Riphah International University, Islamabad, Pakistan. He is currently serving as an Assistant Professor at the Department of Mathematics, Faculty of Arts and Science, Balıkesir University, Balıkesir, Turkey. Before joining Balıkesir University, he worked as a Research Assistant at the Faculty of Engineering of the Free University of Bozen-Bolzano, Italy, and served as an Assistant Professor at Riphah International University in Islamabad, Pakistan. His research interests include computational mathematics, particularly numerical analysis, computational fluid dynamics, iterative methods for nonlinear algebraic and transcendental equations, integral inequalities, and numerical techniques for solving initial and boundary value problems in both classical and fractional frameworks. Dr. Shams has 69 publications, including a book, book chapters, and numerous journal articles in high-impact journals. He has supervised one Ph.D. student, more than 28 MPhil students, and 30 BS Mathematics students. He serves as a reviewer for various international journals and has presented papers at several international conferences as an invited speaker. His significant contributions have advanced the field of computational mathematics and its applications.

Bruno Carpentieri received his Laurea in Applied Mathematics from the University of Bari, Italy, and a Ph.D. in Computer Science from INPT, France. He held postdoctoral positions in Austria and Italy before joining academia. He has been an Assistant Professor at the University of Groningen, a Reader at Nottingham Trent University, and, since 2017, an Associate Professor in Applied Mathematics at the Free University of Bozen-Bolzano. His research focuses on numerical linear algebra, high-performance computing, and computational mathematics, with applications in scientific modeling and engineering. He has served on scientific advisory boards, editorial committees, and review panels for leading journals and conferences. He has supervised over 20 student projects at various levels and authored about 80 peer-reviewed publications.

Contents

Preface

Polynomials are fundamental and widely used structures in mathematics, with deep roots in fields like algebra, calculus, and geometry. Due to their simplicity and versatility, they represent essential tools for solving equations, analyzing data, and modeling real-world phenomena. The structure of polynomials helps us analyze functions and understand their behavior, providing a foundation for many mathematical models in science. In calculus, for example, polynomials provide a natural introduction to topics like differentiation and integration.

This volume explores both classical and modern aspects of the theory and practice of polynomials. It covers recent progress in the development of analytical tools and numerical methods for solving polynomial equations, with applications in science, engineering, and the social sciences. Special attention is given to the numerical behavior of algorithms, including their convergence, stability, and computational efficiency when applied to polynomial systems. Through theory, computations, and case studies, the book helps readers gain a deeper understanding of how algorithms interact with the dynamics of polynomials.

The intended audience comprises advanced undergraduate and graduate students, academic researchers, instructors, and professionals seeking to expand their knowledge of polynomial functions and their practical applications. The book covers both foundational concepts and classifications, as well as advanced topics such as factorization, root-finding methods, interpolation, and symbolic computation. It also includes cutting-edge topics such as polynomial approximations, fractional-order models, and the role of polynomials in computational mathematics.

A distinguishing feature of this book is its dual perspective: it balances the rigor of classical algebraic techniques with insights from contemporary computational approaches. Where appropriate, we incorporate geometric interpretations and real-world examples to underscore the relevance of polynomials in diverse domains.

We want to express our gratitude to the many researchers whose work has enriched the field, and to our colleagues and reviewers for their valuable insights during the preparation of this book. We are also grateful to IntechOpen for providing us with the opportunity to publish this book, and in particular to Karmen Đaleta, the Publishing Process Manager, whose invaluable support and expertise have been instrumental throughout the editorial process of this volume. We hope this book serves as a useful resource and inspiration for all readers interested in the elegance and power of polynomials.

Ultimately, this book highlights how polynomial constructs lie at the heart of mathematical thought and computation. By bridging theory with modern scientific applications, the dialogue underscores the continued significance of polynomials in understanding and solving real-world problems. We hope this combination of tradition and innovation encourages further study and discovery in different fields.

Mudassir Shams
Department of Mathematics,
Faculty of Arts and Science,
Balıkesir University,
Balıkesir, Turkey

Faculty of Engineering,
Free University of Bozen-Bolzano (BZ),
Bozen-Bolzano, Italy

Department of Mathematics and Statistics,
Riphah International University,
Islamabad, Pakistan

Bruno Carpentieri
Faculty of Engineering,
Free University of Bozen-Bolzano,
Bozen-Bolzano, Italy

Chapter 1

Polynomial Points in the Plane of a Triangle

Clark Kimberling

Abstract

Hundreds of named triangle centers have representations in barycentric coordinates that are polynomials in three variables, a, b, c, which may be regarded as the sidelengths of a triangle ABC. Consequently, many central lines (e.g. Euler line, infinity line, Nagel line) also have polynomial representations. Among polynomial representations for points and lines are those of minimal degree. One objective of this chapter is to determine, for selected cases, the minimal degree for triangle centers that lie on a central line, depending on standard forms for polynomial representations of triangle centers. This chapter also discusses polynomial representations of transformations in the plane of ABC, including isotomic conjugate, barycentric product, composition, and symbolic substitution.

Keywords: homogeneous coordinates, barycentric, polynomial, polycenter, polyline, triangle, triangle center, Euler line, infinity line, isogonal conjugate, isotomic conjugate, 2010 *Mathematics Subject Classification*: primary 51 N20; secondary 51 M05

1. Introduction

Among fundamental mathematical expressions in mathematics, and in geometry in particular, are coordinates—not just cartesian coordinates in an xy plane, but also coordinates relative to a fixed triangle. This is especially true for systems known as homogeneous trilinear coordinates (henceforth simply *trilinears* and homogeneous barycentric coordinates (*barycentrics*). These two systems are closely related, and if you you know trilinears for a point or line, you can easily produce barycentrics, and conversly. In the nineteenth century, trilinears were very common in triangle geometry, but in modern triangle geometry, barycentrics prevail. Most of the important points in the plane of a triangle have barycentrics that are polynomials. For example, the first of three barycentrics for the circumcenter of a triangle ABC with sidelengths a, b, c is the polynomial $a^2(b^2 + c^2 - a^2)$. The first of three trilinears for the same point is the polynomial $a(b^2 + c^2 - a^2)$.

Before introducing the notations used in this chapter, we mention two main sources of motivation for the chapter, or, more generally, the ongoing "transfiguration" of the subject as predicted by Philip J. Davis [1] some 30 years ago: powerful computer programs, such as *Mathematica*, and the recognition that virtually every object in triangle geometry (points, lines, curves, etc.) has a formula, as hinted in the preceding paragraph.

Where in the literature can one track the rise of modern triangle geometry? Davis [1] has an excellent bibliography—up to 1992. For more recent developments, see the combined volumes of *Forum Geometricorum* [3] and the articles by Sava Grosdev, Dao Thanh Oai, Stanley Rabinowitz, and Paul Yiu in the open-access *International Journal of Computer Discovered Mathematics* [6]. Other articles that use polynomial representations in triangle geometry deal with poristic families of triangle centers; see [12, 14].

Following Leonhard Euler [15], we label a triangle ABC with corresponding sidelengths a, b, c as in **Figure 1**. Then, following Augustus Ferdinand Möbius (1827), each point P in the plane of ABC has barycentrics determined as follows. A point P is a vertex of (possibly degenerate) triangles BPC, CPA, APB, whose areas we denote by $\sigma_{BPC}, \sigma_{CPA}, \sigma_{APB}$. If p, q, r are any three numbers respectively proportional to the three *signed* areas, then $p : q : r$ are barycentrics for P. (Regarding the modifier "signed", if, for example, P lies on the side of line BC that does not include vertex A, then the signed area of triangle BPC is $-\sigma_{BPC}$.) For details and an overall excellent introduction to barycentric coordinates and their role in modern triangle geometry, see Paul Yiu's online book [16].

For P in the interior of ABC, **Figure 1** shows barycentrics for the points A, B, C, and P.

As an example of a triangle center—and of remarkable flexibility when dealing with homogeneous coordinates—the circumcenter, O, has barycentrics as follows:

$$O = \sin 2A : \sin 2B : \sin 2C \tag{1}$$

$$\begin{aligned}
&= 2\sin A\cos A : 2\sin B\cos B : 2\sin C\cos C\\
&= a\cos A : b\cos B : c\cos C\\
&= a(b^2+c^2-a^2)/(2bc) : b(c^2+a^2-b^2)/(2ca) : c(a^2+b^2-c^2)/(2ab)\\
&= a^2(b^2+c^2-a^2) : b^2(c^2+a^2-b^2) : c^2(a^2+b^2-c^2)
\end{aligned} \tag{2}$$

The first barycentric in (2) is a polynomial in a, b, c. We define a *polynomial center* (or simply *polycenter*) to be a triangle center [5], X, if there exists a polynomial $x(a, b, c)$, all of whose coefficients are integers, such that

$$X = x(a,b,c) : x(b,c,a) : x(c,a,b). \tag{3}$$

Thus, O is a polycenter. The definition of triangle center [5] implies that the polynomial $x(a, b, c)$ is homogeneous in a, b, c, and $x(a, c, b) = x(a, b, c)$. It is often

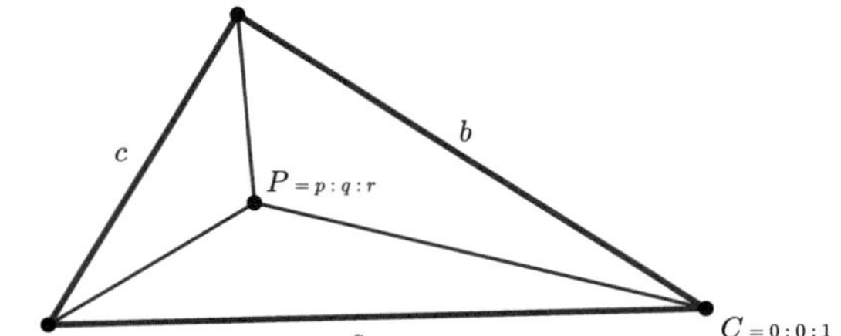

Figure 1.
Triangle ABC and point P with barycentric coodinates p : q : r.

convenient to write simply $x(a,b,c)$ :: for X, since the first barycentric determines the other two, as in (3).

Most named triangle centers are polycenters, as illustrated by the first eleven such points indexed in the Encyclopedia of Triangle Centers (ETC) [2]. These triangle centers appear in **Table 1**.

A few details regarding **Table 1** may be enlightening. The first four polycenters, indexed as X_1, X_2, X_3, X_4, are known as the ancient Greek points; they appear in Euclid's *Elements*. The nine-point center, X_5, is the center of the nine-point circle, which is tangent to the incircle at the Feuerbach point, X_{11}. The eleven points in **Table 1** all have individual entries in both Mathworld [9] and Wikipedia.

1.1 Polynomial lines (polylines)

The line passing through two distinct triangle centers is called a central line, given by an equation

$$x\alpha + y\beta + z\gamma = 0, \tag{4}$$

where $\alpha : \beta : \gamma$ is a variable point and $x : y : z$ is a triangle center. The line (4) is often abbreviated as $[x : y : z]$, or simply $[x]$. If $x : y : z$ is a polycenter, the line (4) is a *polyline*. Three of the most important lines in triangle geometry are polylines:

1. infinity line, $[1 : 1 : 1]$, denoted by L^{∞};

2. Euler line, $\left[\left(b^2 - c^2\right)\left(a^2 - b^2 - c^2\right)\right]$;

3. Nagel line, $[b - c]$.

For a list of central lines, most of which are polylines, see Ref. [10].

Name	Index in ETC	First barycentric
Incenter	X_1	a
Centroid	X_2	1
Circumcenter	X_3	$a^2(a^2 - b^2 - c^2)$
Orthocenter	X_4	$\left(b^2 - c^2 - a^2\right)\left(c^2 - a^2 - b^2\right)$
Nine-point center	X_5	$a^2\left(b^2 + c^2\right) - \left(b^2 - c^2\right)^2$
Symmedian point	X_6	a^2
Gergonne point	X_7	$(a - b + c)(a + b - c)$
Nagel point	X_8	$b + c - a$
Mittenpunkt	X_9	$a(b + c - a)$
Center of Spieker circle	X_{10}	$b + c$
Feuerbach point	X_{11}	$(b + c - a)(b - c)^2$

Table 1.
Polycenters with polynomial first barycentric.

1.2 Standard form

The first barycentric of a polycenter is a linear combination, with integer coefficients, of polynomials of the form

$$a^m\left(b^ic^j+b^jc^i\right), \tag{5}$$

where $m\geq 0, i\geq 0, j\geq 0$. For example, the polycenter $b-c$ :: is given by

$$\begin{aligned} b-c ::&= (b-c)^2(a-b)(a-c) :: \\ &= a^2\left(b^2+c^2\right)-2a^2bc-a\left(b^3+c^3\right) \\ &+a\left(b^2c+bc^2\right)+b^3c+bc^3-2b^2c^2 :: . \end{aligned}$$

This example shows that a polycenter can be represented one way using a polynomial of degree 1 and another way with a polynomial of degree 4. Generally, of all possible polynomial representations for a polycenter X, there is a minimal possible degree, which we use to define the *degree* of the polycenter.

If

$$X = x(a,b,c) : x(b,c,a) : x(c,a,b) \tag{6}$$

is a representation of minimal degree and $x(a,b,b)=0$, then X is an *odd polycenter*; otherwise X is an *even polycenter*. Clearly, an odd polycenter $x(a,b,c)$ :: has the form $(b-c)x^*(a,b,c)$ ::, where $x^*(a,b,c)$ :: is an even polycenter that is not a multiple of $(a-b)(a-c)$. Likewise, we can refer to the degree and parity of a polyline $[x(a,b,c)]$.

To summarize, a polycenter (6) (or polyline $[x]$) has a *standard form* if $x(a,b,c)$ has minimal degree (call it N) and the form

$$x(a,b,c)=\sum_{m=0}^{n}k_m a^m h_m(b,c), \tag{7}$$

where

1. $k_n>0$;

2. the coefficients k_m are coprime integers;

3. $h_m(b,c)$ is of the form $b^ic^j+b^jc^i$ for $m=0,1\ldots,n, i\geq 0, j\geq 0$;

4. the degree of $h_m(b,c)$ is $N-m$.

1.3 Polycenters on polylines

Here, we are interested in the following sort of question: *what is the minimal degree of polycenter that lies on the Euler line?* We begin by recognizing general forms for first barycentrics $x(a,b,c)$ of polycenters by degree and parity:

degree 0: 1.

degree 1:

even: $h_1a+h_2(b+c)$

odd: $b-c$

degree 2:

even: $h_1a^2 + h_2a(bc) + h_3bc + h_4(b^2 + c^2)$
odd: $(b - c)(h_1a + h_2(b + c))$
degree 3:
even: $h_1a^3 + h_2a^2(b + c) + h_3abc + h_4a(b^2 + c^2) + h_5(bc^2 + b^2c) + h6(b^3 + c^3)$
odd: $(b - c)(h_1a^2 + h_2a(bc) + h_3bc + h_4(b^2 + c^2))$
degree 4:
even: $h_1a^4 + h_2a^3(b + c) + h_3a^2bc + h_4a^2(b^2 + c^2) + h_5a(b^2c + bc^2) + h_6(b^3c + bc^3) + h_7(b^3c + bc^3) + h_8b^2c^2 + h_9(b^4 + c^4)$
odd: $(b - c)(h_1a^3 + h_2a^2(b + c) + h_3abc + h_4a(b^2 + c^2) + h_5(bc^2 + b^2c) + h6(b^3 + c^3)$

Next, we consider a systematic way to describe individual terms $k_m a^m h_m(b, c)$ as in (7). Let $f(N)$ be the number of such terms, so that $f(0) = 1$, there being only one polycenter of degree 0. Assume now that $N \geq 1$ and that X is an even polycenter. There are two kinds of terms, for which we have cases:

Case 1: $m > 0$. Here, each term is of the form aT, where T is one of the $f(N - 1)$ terms comprising (7) when the degree is $N - 1$.

Case 2: $m = 0$. If N is even, we write $N = 2w$ and note that the sums $b^ic^j + b^jc^i$ in (5) are these $w + 1$ sums:

$$b^{2w} + c^{2w}, b^{2w-1}c + bc^{2w-1}, \ldots, b^wc^w.$$

If N is odd, we write $N = 2w - 1$ and count these w sums:

$$b^{2w-1} + c^{2w-1}, b^{2w-2}c + bc^{2w-2}, \ldots, b^wc^{w-1} + b^{w-1}c^w.$$

The two cases account for a total of

$$f(N) = f(N - 1) + \lfloor N + 2)/2 \rfloor$$

terms. The sequence $(f(N))$ is essentially $A002620$ in [13].

With the aid of a computer, we can now prove, for example, that no polycenter of degree 1 lies on the Euler line. Mathematica coding for such purposes is typified by the following code, which is used to determine the form of every 5th degree polycenter that lies on the Euler line.

```
p[ a_, b_, c_] := (b^2 - c^2) (a^2 - b^2 - c^2);
f[ a_, b_, c_] :={ p[ a, b, c] , p[ b, c, a] , p[ c, a, b] } ;
    (* f: Euler line *)
u[ a_, b_, c_] := h1* a^5 + h2* a^4 (b + c) + h3* a^3* b* c
    + h4* a^3* (b^2 + c^2)
    + h5* a^2* (b^3 + c^3) + h6* a* (b^2* c + b* c^2)
    + h7* a* (b^3* c + b* c^3) + h8* a* b^2* c^2
    + h9* a* (b^4 + c^4) + h10* (b^5 + c^5)
    + h11* (b^4* c + b* c^4) + h12 (b^3* c^2 + c^2* c^3);
g[ a_, b_, c_] ={ u[ a, b, c] , u[ b, c, a] , u[ c, a, b] } ;
    (* g: arbitrary 5th degree even polycenter *)
w = Factor[ f[ a, b, c] . g[ a, b, c] ]
    (* w=Factor[ w/ ( (b-c) (c-a) (a-b) )] *)
cc = CoefficientList[ w, a] ;
z = ConstantArray[ 0, Length[ cc] ] ;
Solve[ cc == z, { h1, h2, h3, h4, h5, h6, h7, h8, h9, h10, h11, h12} ]
```

```
    (* Got{ h4 = h11-h2, h5 = h11-h2, h6 = 0, h7 = h3,
h8 = -h11+h2, h9 = h11, h10 = h1+h11-h2, h12 =0} *)
uu = h1* a^5 + h2* a^4 (b + c) + h3* a^3* b* c +
(h11 - h2)* a^3* (b^2 + c^2) + (h11 - h2)* a^2* (b^3 + c^3)
+ h3* a* (b^3* c + b* c^3) + (-h11 + h2)* a* b^2* c^2 + h11* a* (b^4 + c^4)
+ (h1 + h11 - h2)* (b^5 + c^5) + h11* (b^4* c + b* c^4);
Collect[ uu, { h1, h2, h3, h11} ]
```

To see how the code works, and how it can be used for solving problems involving other lines and points, note that the coefficients for the line are given by

`f[ a_,b_,c_]`

and that the general form for a polycenter (here of degree 5) is given by

`u[ a_,b_,c_]`

The key line of code uses

`Solve`

to solve the system of linear equations for the equation $w = 0$, this system being equivalent to the polycenter lying on the polyline. The remark, beginning with

`Got`

must be supplied to the code from the solution of the system. The last line of code shows the independent coefficients in $u[a, b, c]$ for which the polycenter $g(a, b, c)$ :: lies on the given line. The associated basis-polycenters are linearly independent, which is to say that the others initially built into $u(a, b, c)$ are linear combinations of the basis-polycenters. The associated polycenters comprise a basis for all polycenters of the chosen degree of $u(a, b, c)$.

The next three sections summarize of results obtained from variations of the above quoted code, for three specific lines: Euler line, infinity line, and Nagel line. These results are of two possible kinds: (1) no polycenters of a particular parity and degree on the line in question; or (2), a linear combination of polycenters. In the latter case, the first barycentric of such a polycenter has the form

$$\sum_{m=1}^{n} k_i g_i(a, b, c),$$

where the coefficients k_i are real numbers, not all 0, and the polynomial $g_i(a, b, c)$ has the form $s(a, b, c)p(a, b, c)$, where $s(a, b, c)$ is a symmetric polynomial and $p(a, b, c)$ is the first barycentric of a polycenter in standard form.

The three lines just mentioned were chosen for this chapter because of their ubiquity and simplicity, not because of any particular connections among them (although there are interesting polynomial-type mappings from any line onto any other line). Also, as is the case with most discussions of plane geometry, there are no well-known practical applications of these lines. One can, however, make the following observation about the Euler and Nagel lines as they cross a horizontal physical triangle having constant mass density: that the triangle will balance on top of a vertical knife-edge that coincides with either of the two special lines. This property stems from the fact that the centroid of the triangle lies on both lines.

DOI: http://dx.doi.org/10.5772/intechopen.1007000

2. Polycenters on the Euler line

More than 6000 triangle centers on the Euler line, most of which are polycenters, are identified in [2, 10]. Among them are the ancient Greek points: centroid, $1:1:1$, circumcenter, $a^2(a^2-b^2-c^2)$ ::, and orthocenter, $1/(a^2-b^2-c^2)$ ::.

The code quoted above can be readily adapted to prove that only one polycenter of degree less than 4 lies on the Euler line, and that is the centroid. The code was used to find that no odd polycenter of degree ≤ 7 lies on the Euler line and also, to give the following results for even polycenters of degrees 4, 5, 6, 7 on the Euler line.

The general even 4*th* degree polycenter on the Euler line has first barycentric $\sum_{i=0}^{5} k_i g_i(a,b,c)$, where

$$\begin{aligned}
g_1(a,b,c) &= a^4+b^4+c^4;\\
g_2(a,b,c) &= abc(a+b+c);\\
g_3(a,b,c) &= a^3(b+c)+b^3(c+a)+c^3(a+b);\\
g_4(a,b,c) &= a^2(b^2+c^2)+b^4+c^4;\\
g_5(a,b,c) &= b^2c^2-b^4-c^4,
\end{aligned}$$

where $g_i(a,b,c)$ :: is the centroid for $i=1{,}2{,}3$, whereas $g_4(a,b,c)=X_{6656}$ and $g_5(a,b,c)=X_{5025}$.

The general even 5*th* degree polycenter on the Euler line has first barycentric $\sum_{i=0}^{4} k_i g_i(a,b,c)$, where

$$\begin{aligned}
g_1(a,b,c) &= a^5+b^5+c^5;\\
g_2(a,b,c) &= abc(a^2+b^2+c^2);\\
g_3(a,b,c) &= a^4(b+c)-a^3(b^2+c^2)-a^2(b^3+c^3)+ab^2c^2-b^5-c^5;\\
g_4(a,b,c) &= a^3(b^2+c^2)+a^2(b^3+c^3)+a(b^4+c^4-b^2c^2)+b^5+c^5+b^4c+bc^4.
\end{aligned}$$

Here, $g_i(a,b,c)$ :: is the centroid for $i=1,2$, whereas $g_3(a,b,c)=X_{63795}$ and $g_5(a,b,c)=X_{63794}$.

The general even 6*th* degree polycenter on the Euler line has first barycentric $\sum_{i=0}^{7} k_i g_i(a,b,c)$, where

$$\begin{aligned}
g_1(a,b,c) &= a^6+b^6+c^6;\\
g_2(a,b,c) &= abc(a^3+b^3+c^3);\\
g_3(a,b,c) &= a^2b^2c^2;\\
g_4(a,b,c) &= b^3c^3+c^3a^3+a^3b^3;\\
g_5(a,b,c) &= b^6+c^6+a^2(b^4+c^4);\\
g_6(a,b,c) &= (bc+ca+ab)(b^4+c^4-b^2c^2);\\
g_7(a,b,c) &= a^5(b+c)-a(b^4c+bc^4-b^3c^2-b^2c^3)-b^3c^3.
\end{aligned}$$

Here, $g_i(a,b,c)$ :: is the centroid for $i=1{,}2{,}3{,}4$. Note that although $g_6(a,b,c)$ is a polynomial of degree 6, the corresponding polycenter, $b^4+c^4-b^2c^2$ ::, has degree 4.

The general even *7th* degree polycenter on the Euler line has first barycentric $\sum_{i=0}^{10} k_i g_i(a,b,c)$, where

$$g_1(a,b,c) = a^7 + b^7 + c^7;$$
$$g_2(a,b,c) = abc(b^2c^2 + c^2a^2 + a^2b^2);$$
$$g_3(a,b,c) = a^2b^2c^2(a + b + c);$$
$$g_4(a,b,c) = b^4c^3 + b^3c^4 + c^4a^3 + c^3a^4 + a^4b^3 + a^3b^4;$$
$$g_5(a,b,c) = abc(b^4 + c^4 - b^2c^2));$$
$$g_6(a,b,c) = abc(a^4 + b^2c^2);$$
$$g_7(a,b,c) = b^7 + c^7 + a^5(b^2 + c^2) + a^4(-b^3 - c^3) + a^2(b^5 + b^3c^2 + b^2c^3 + c^5);$$
$$g_8(a,b,c) = b^6c + bc^6 - a^5(b^2 + c^2) + a^4(b^3 + c^3) + a^3(b^4 + c^4) + a^2(b^4c + bc^4) + a\left(b^6 + c^6\right);$$
$$g_9(a,b,c) = -b^7 + b^5c^2 + b^2c^5 - c^7 + a^4(b^3 + c^3) + a^2(b^3c^2 + b^2c^3 - b^4c - bc^4);$$
$$g_{10}(a,b,c) = a^6(b + c) + a^5(b^2 + c^2) - a^4(b^3 + c^3) + a^2(b^3c^2 + b^2c^3)$$
$$-a^3(b^4 + c^4) - a^2(b^4c + bc^4).$$

Here, $g_i(a,b,c)$ :: is the centroid for $i = 1,2,3,4$. The point $g_6(a,b,c)$ :: is the celebrated Conway point, $X_{384} = a^4 + bg^2c^2$ ::, of degree 4, whereas the basis-polynomial, $abc(a^4 + b^2c^2)$, has degree 7.

3. Polycenters on the infinity line

The infinity line, denoted by L^∞, consists of points X whose barycentric coordinates $x : y : z$ satisfy

$$x + y + z = 0. \tag{8}$$

Algebraic and geometric properties of such points are exemplied [2] by the infinite point of the Euler line, X_{30}, and X_i for $i = 511,512, \dots, 545$.

There is exactly one odd polycenter of degree 1 on L^∞, and that is $X_{514} = b - c$ ::.

A proof that every odd *2nd* degree polycenter lies on L^∞ follows: a general formula for these points is $(b - c)(ha + k(b + c))$ ::, where h and k are unequal real numbers, and we have

$$(b - c)(ha + k(b + c)) + (c - a)(hb + k(c + a) + (a - b)(hc + k(a + b)$$
$$= h(a(b - c) + b(c - a) + c(a - b)) + k((b^2 - c^2) + (c^2 - a^2) + (a^2 - b^2)) = 0.$$

Adaptations of the Mathematica code quoted above yield the following general forms, by first barcentric, for polycenters on L^∞.

Even 2nd degree: $k_1(2a^2 - b^2 - c^2) + k_2(-2bc + ca + ab))$. That is, the polycenters X_{524} and X_{536} comprise a basis for all such polycenters.

Odd 3rd degree: $k_1(b - c)(a^2 - bc) + k_2(b - c)(b^2 + c^2 + bc)$. Here the basis consists of X_{812} and X_{824}.

Even 3rd degree: $k_1(b + c)(-a^2 + bc) + k_2(a^2b^2 + a^2c^2 - b^2c - bc^2)$. Basis: X_{740} and X_{726}.

Odd 4th degree polycenter on L^{∞}, with basis-polycenters indicated by 1st barycentric:

$$\begin{array}{lr} k_1(b-c)\left(a^3-b^2-bc^2\right) & X_{814} \\ +k_2a^2\left(b^2-c^2\right) & X_{512} \\ +k_3abc(b-c) & abcX_{514} \\ +k_4(b-c)\left(ab^2+ac^2+b^2c+bc^2\right) & X_{63812} \end{array}$$

Even 4th degree polycenter on L^{∞}, with basis polycenters indicated by 1st barycentric:

$$\begin{array}{ll} k_1\left(2a^4-b^4-c^4\right) & X_{754} \\ +k_2\left(a^3b+a^3c-b^3c-b^3\right) & X_{742} \\ +k_3\left(2a^2bc-ab^2c-abc^2\right) & X_{519} \\ +k_4\left(-2b^2c^2+a^2b^2+a^2c^2\right) & X_{538} \\ +k_5\left(ab^3+ac^3-b^3c-bc^3\right) & X_{712} \end{array}$$

Odd 5th degree polycenter on L^{∞}, with basis polycenters indicated by 1st barycentric:

$$\begin{array}{lr} k_1(b-c)\left(a^4++b^4+c^4\right) & \left(a^4+b^4+c^4\right)X_{514} \\ +k_2(b-c)\left(a^3b+a^3c-b^2-c^2\right) & X_{802} \\ +k_3a^2bc(b-c) & abcX_{513} \\ +k_4(b-c)\left(b^2c^2+c^2a^2+a^2b^2\right) & \left(b^2c^2+c^2a^2+a^2b^2\right)X_{514} \\ +k_5(b-c)\left(b^4+c^4-ab^3-ac^3\right) & X_{63813} \\ +k_6abc\left(b^2-c^2\right) & abcX_{523} \\ +k_7(b-c)\left(b^4+c^4+b^3c+bc^3+b^2c^2\right) & X_{63814} \end{array}$$

Even 5th degree polycenter on L^{∞}, with basis polycenters indicated by 1st barycentric:

$$\begin{array}{lr} k_1\left(2a^5-b^5-c^5\right) & X_{33911} \\ +k_2(b+c)\left(a^4-b^3c-bc^3+b^2c^2\right) & X_{744} \\ +k_3abc\left(2a^2-b^2-c^2\right) & abcX_{524} \\ +k_4a^2\left(b^3+c^3-ab^2-ac^2\right) & X_{674} \\ +k_5(ab^4+ac^4-b^4c-bc^4 & X_{696} \end{array}$$

4. Polycenters on the Nagel line

Unlike the Euler line and the infinity line, the Nagel line, given by

$$(b-c)x+(c-a)y+(a-b)z=0,$$

is an odd line.

Every even 2nd degree polycenter has the form $ha + k(b + c)$ ::, and it is easy to check that every such point lies on the Nagel line.

Adaptations of the Mathematica code quoted above yield the following general forms, by first barcentric, for polycenters on the Nagel line:

Even 3rd degree: $k_1(a^3 - b^2c - bc^2) + k_1a^2(b + c) + k_3abc + k_4(a(b^2 + c^2)$

Basis: $\{X_{4362}, X_{42}, X_{38}\}$.

Even 4th degree polycenters on the Nagel line, with basis polycenters indicated by first barycentric:

$k_1(a^4 + b^4 + c^4)$	$(a^4 + b^4 + c^4)X_2$
$+k_2(a^3b + a^3c - b^2c^2)$	X_{17033}
$+k_3a^2bc$	$abcX_1$
$+k_4(b^2c^2 + c^2a^2 + a^2b^2)$	$(b^2c^2 + c^2a^2 + a^2b^2)X_2$
$+k_5(b^4 + c^4 - ab^3 - ac^3)$	X_{63817}
$+k_6abc(b + c)$	$abcX_{10}$
$+k_7b^4 + c^4 + b^3c + bc^3 + b^2c^2$	X_{30149}

Odd 5th degree polycenters on the Nagel line, with basis polycenters indicated by first barycentric:

$k_1(a^5 + b^5 + c^5)$	X_{27486}
$+k_2(b + c)(a^4 + b^4 + c^4))(a^4 + b^4 + c^4)$	X_{10}
$+k_3(abc(a^2 - bc))$	X_{239}

Even 5th degree polycenters on the Nagel line, with basis polycenters indicated by first barycentric:

$k_1(a^5 + b^5 + c^5)$	X_{27486}
$+k_2(b + c)(a^4 + b^4 + c^4))$	$(a^4 + b^4 + c^4)X_{10}$
$+k_3(abc(a^2 - bc))$	X_{239}
$+k_4a(b^2c^2 + c^2a^2 + a^2b^2)$	$(b^2c^2 + c^2a^2 + a^2b^2)X_1$
$+k_5a(a^3(b + c) + a(b^3 + c^3) - b^2c^2)$	X_{63818}
$+k_6abc(b + c)$	X_{3661}
$+k_7abc(b^2 + bc + c^2))$	X_{3661}
$+k_8(b^5 + c^5 - ab^4 - ac^4)$	X_{63819}

5. Algebraic structure among polycenters and polylines

In this section, we begin with what can be considered foremost connection between triangle algebra and triangle geometry. Suppose that

$$P_1 = x_1 : y_1 : z_1, \quad P_2 = x_2 : y_2 : z_2, \quad P_3 = x_3 : y_3 : z_3$$

are points, and let D be the determinant

$$\begin{vmatrix} x_1 & y_1 & z_1 \\ x_2 & y_2 & z_2 \\ x_3 & y_3 & z_3 \end{vmatrix}$$

Then P_1, P_2, P_3 are collinear if and only if $D = 0$. Dually, if three lines L_1, L_2, L_3 are given by

$$\begin{aligned} x_1\alpha + y_1\beta + z_1\gamma &= 0 \\ x_2\alpha + y_2\beta + z_2\gamma &= 0 \\ x_3\alpha + y_3\beta + z_3\gamma &= 0, \end{aligned}$$

then L_1, L_2, L_3 concur in a single point if and only if $D = 0$.

Although this fundamental duality is well known, there are many other algebraic, and hence polynomial-represented, aspects of triangle geometry. In most of the literature, the focus is on geometric features, rather than algebraic. A good example is the isotomic conjugate of a point $P = p : q : r$ in the plane of a triangle ABC, and not on any of the sidelines BC, CA, AB. The geometric construction goes like this: let M_A, M_B, M_C be the respective midpoints of the sides BC, CA, AB. Let L_A be the reflection of line AP in the line AM_A, and likewise define lines L_B and L_C. The lines L_A, L_B, L_C concur in a single point—denote it by P^{-1}—which is the isotomic conjugate of P. Now, here is the algebraic version: $P^{-1} = 1/p : 1/q : 1/r$.

With the foregoing example in mind, let $\mathcal{P}$ denote the set of polycenters (i.e., triangle centers whose first barycentric coordinate is a polynomial in a, b, c, assumed to be in standard form whenever this is necessary to avoid ambiguity.) Clearly, $(\mathcal{P}^*)$ *is a commutative group under barycentric multiplication,* *. In particular, the identity is the centroid, $1 : 1 : 1$, and the inverse of a polycenter $P = p(a,b,c)$ :: is $1/P = 1/p(a,b,c)$ ::, this being the isotomic conjugate of P, which in polynomial form is $p(b,c,a) * p(c,a,b)$ ::.

Surprisingly, barycentric multiplication is relatively new to the literature. For a construction and discussion, see Yiu ([16], pp. 99–101).

5.1 Trace, combo, and midpoint

In addition to *, there are many other operations (i.e., transformations, conjugacies, inverses, etc.) under which $\mathcal{P}$ is closed. Among them is a form of addition known as a combo. To define it, we shall call the sum $p(a,b,c) + p(b,c,a) + p(c,a,b)$ the *trace* of P, denoted by $tr(P)$. This function, symmetric in a, b, c, is used to define the *normalized barycentrics* of a point $X = (x, y, z)$ satisfying $x + y + z \neq 0$ (i.e., X is not on the infinity line) as the ordered triple

$$N(X) = (x', y', z') = (x/tr(P), y/tr(P), z/tr(P)).$$

Now suppose that P and U are two such polycenters, and that $h = h(a,b,c)$ and $k = k(a,b,c)$ are symmetric in (a,b,c). The (h,k)*-combo* of P and U is the point $hN(P) + kN(U)$ ::, which is essentially a linear combination P and U. For example, geometrically, the $(1/2, 1/2)$-combo of P and U is the midpoint of P and U. Among the most commonly encountered combos are those associated with polycenters on the Euler line. Several examples can be read from the equations

$$X_5 = 3^* X_2 + X_4 = 3^* X_2 - X_3 = X_3 + X_4.$$

(The polycenters represented here are the nine-point center X_5, centroid X_2, orthocenter X_4, and circumcenter X_3.) For more about combos, see the Introduction of ETC [2].

Returning to traces, note that a point P lies on the infinity line if and only if $tr(P) = 0$. Equivalently, a line given by $\ell x + my + nz = 0$ is the infinity line if and only if $tr(\ell : m : n) = 0$. We turn now to selected examples of functions f symmetric in a, b, c and polycenters having f as trace.

Every even 1*st* degree polycenter has the form $h_1 a + h_2(b + c)$, so that its trace is

$$h_1(a + b + c) + h_2(2a + 2b + 2c).$$

This shows that every 1*st* degree polycenter has trace of the form $k(a + b + c)$.

Every odd 3*rd* degree polycenter has the form

$$(b - c)\left(h_1 a^2 + h_2 a(b + c) + h_3 bc + h_4\left(b^2 + c^2\right)\right), \tag{9}$$

from which it is easy to see that every such polycenter has trace of the form $k(b - c)(c - a)(a - b)$, where k is a nonzero real number, and that no even 3*rd* degree polycenter has that trace.

The next of our selected symmetric functions is

$$\begin{aligned} &k(a + b + c)(b + c - a)(c + a - b)(a + b - c) \\ &= k\left(2b^2c^2 + 2c^2a^2 + 2a^2b^2 - a^4 - b^4 - c^4\right), \end{aligned} \tag{10}$$

this being 16k times the square of the area of the reference triangle ABC. Recalling from Section 1.3 that every even 4*th* degree polycenter has first barycentric of the form

$$\begin{aligned} &h_1 a^4 + h_2 a^3(b + c) + h_3 a^2 bc + h_4 a^2\left(b^2 + c^2\right) + h_5 a\left(b^2 c + bc^2\right) \\ &\quad + h_6\left(b^3 c + bc^3\right) + h_7\left(b^3 c + bc^3\right) + h_8 b^2 c^2 + h_9\left(b^4 + c^4\right), \end{aligned}$$

we find the trace to be

$$\begin{aligned} &(h_1 + 2h_9)\left(a^4 + b^4 + c^4\right) + (h_2 + h_6 + h_7)\left(a^3 b + a^3 c + b^3 c + b^3 a + c^3 a + ca^3\right) \\ &+ (h_3 + 2h_5)abc(a + b + c) + (2h_4 + h_8)\left(b^2c^2 + c^2a^2 + a^2b^2\right). \end{aligned} \tag{11}$$

Equating (10) and (11) and putting

$$h_1 = k - 2h_9, \quad h_8 = -2k - 2h_4, \quad h_9 = r, \quad \text{and } h_4 = s,$$

we obtain the following first barycentric for the family of all 4*th* degree polycenters having trace $k(a + b + c)(b + c - a)(c + a - b)(a + b - c)$:

$$(k - 2r)a^4 + sa^2\left(b^2 + c^2 + (-2k - 2s)b^2c^2 + r\left(b^4 + c^4\right)\right. \tag{12}$$

Among triangle centers having trace (12) are these points in ETC [2]: X_i for $i =$ 63922 to 63957. There is a remarkable affinity between this trace and the Euler line. To see this, let

$$p(a,b,c) = (b^2 - c^2)(a^2 - b^2 - c^2) \tag{13}$$

$$f(a,b,c) = (k - 2r)a^4 + sa^2(b^2 + c^2) + (-2k - 2s)b^2c^2 + r(b^4 + c^4). \tag{14}$$

Then the solution of the equation

$$p(a,b,c)f(a,b,c) + p(b,c,a)f(b,c,a) + p(c,a,b)f(c,a,b) = 0$$

is $k = r - s$, so that the desired points are

$$r(a^4 - b^4 - c^4 + 2b^2c^2) + sa^2(a^2 - b^2 - c^2) :: , \tag{15}$$

or, equivalently, all combos of the orthocenter and circumcenter in which the coefficients, r and s, respectively, are real numbers, not both 0. (If r and s are allowed to be nonconstant homogeneous functions symmetric in a, b, c, then *all* the other points on the Euler line are represented by (15)—but these have degrees other than 4.) For various choices of *integers* r and s, the combo (15) represents these well-known points on the Euler line: $X_3, X_4, X_5, X_{20}, X_{30}, X_{382}$.

The procedure used just above for identifying all even 4*th* degree triangle centers on the Euler line can be used for other degrees and lines. One more result of this sort is that every 4*th* degree triangle center on the line at infinity is a member of the family

$$r(2a^4 - b^4 - c^4) + s(2b^2c^2 - a^2b^2 - a^2c^2) :: ,$$

where r and s are real numbers, not both 0.

5.2 Conjugates and curves

In the group (P, $*$), the multiplicative inverse of a polycenter $P = p(a,b,c)$ :: is the isotomic conjugate of P, given by

$$1/P = 1/p(a,b,c) ::= p(b,c,a) * p(c,a,b) :: \tag{16}$$

Because isotomic conjugacy is a function, we may refer to the isotomic conjugate (alias simply *inverse*) of lines and other sets of points. Of particular interest is the inverse of L^∞, given by the equation $1/x + 1/y + 1/z = 0$, or equivalently, by $yz + zx + xy = 0$. This curve is the Steiner circumellipse. More generally, as is well-known, the inverse of a polyline

$$f(a,b,c)x + f(b,c,a)y + f(c,a,b)z = 0$$

is the circumconic (i.e., a conic that passes through the vertices A, B, C) given by

$$f(b,c,a)f(c,a,b)yz + f(c,a,b)f(a,b,c)zx + f(a,b,c)f(b,c,a)xy = 0.$$

Among all circumconics, perhaps the most familiar is the circumcircle, which is the isogonal (not isotomic) conjugate of L^∞, given by

$$a^2yz + b^2zx + c^2xy = 0.$$

To summarize, if L is a polyline, then both its isotomic and isogonal conjugates are circumconics whose coefficients form a polycenter. Similar representations abound in barycentric and trilinear representations of cubics and higher-order curves [4].

6. Composition

A note dated July, 2024, in ([2], Part 33, preamble just before X_{64425}) defines a composition operation on the set of polycenters, as follows. Suppose that $P = p(a,b,c) : p(b,c,a) : p(c,a,b)$ and $U = u(a,b,c) : u(b,c,a) : u(c,a,b)$ are triangle centers, where $p(a,b,c)$ and $u(a,b,c)$ are polynomials in standard form (i.e., $p(a,b,c)$ and $p(b,c,a)$ are relatively prime, and the coefficient of the highest power of a is positive, or if $p(a,b,c)$ is invariant of a then the coefficient of highest power of b is positive). Define the composite, denoted by P-of-U, to be the triangle center given by

$$\begin{aligned} P\text{-of-}U &= p(u(a,b,c),u(b,c,a),u(c,a,b)) \\ &: p(u(b,c,a),u(c,a,b),u(a,b,c)) \\ &: p(u(c,a,b),u(a,b,c),u(b,c,a)). \end{aligned}$$

If the degrees of P and U are m and n, then the degree of P-of-U is clearly at most mn. That the degree of P-of-U can be less than mn is exemplified by the fact that P-of-$X_2 = X_2$ for every polycenter P, regardless of the degree of P. A less trivial example is X_4-of-$X_3 = X_{68}$; here, the degrees of X_4, X_3, and X_{68} are 4, 4, and 10, respectively.

7. Symbolic substitution

Suppose that $P = p(a,b,c) : p(b,c,a) : p(c,a,b)$ and $U = u(a,b,c) : u(b,c,a) : u(c,a,b)$ are polycenters. Then the symbolic substitution [7] denoted by $(a,b,c) \to (u(a,b,c),u(b,c,a),u(c,a,b))$ maps P onto the polycenter

$$p(u(a,b,c),u(b,c,a),u(c,a,b) :: . \tag{17}$$

Symbolic substitutions of this kind, which map polycenters to polycenters, polylines to polylines, and likewise for higher-degree curves, are discussed in [7].

Regarded as strictly algebraic operations, symbolic substitutions are natural and amenable; interestingly, however, there seem to be no corresponding general geometric constructions. The algebraic and geometric aspects of symbolic substitutions are illustrated by **Table 2**, which consists of the polycenters obtained by applying the symbolic substitution $\sigma : a \to bc$ to the eleven polycenters in **Table 1**.

Similarly, **Table 3** shows the results of applying the symbolic substitution $\tau : a \to b - c$ to the same eleven polycenters.

Certain differences between **Tables 2** and **3** are striking. In particular, only **Table 3** has distinct polycenters mapped by τ to the same point (e.g., $\tau(X_6) = \tau(X_9) = X_{1086}$), and only in **Table 3** do we see symbolic substitutions on the line at infinity, L^∞. One might ask: how does τ treat the polycenters on L^∞? One answer is given by the following theorem.

name	index in ETC	first barycentric
σ(incenter)	$\sigma(X_1) = X_{75}$	bc
σ(centroid)	$\sigma(X_2) = X_2$	1
σ(circumcenter)	$\sigma(X_3) = X_{6374}$	$b^2c^2(b^2c^2 - c^2a^2 - a^2b^2)$
σ(orthocenter)	$\sigma(X_4) = X_{2998}$	$(c^2a^2 - a^2b^2 - b^2c^2)(a^2b^2 - b^2c^2 - c^2a^2)$
σ(nine-point center)	$\sigma(X_5) = X_{6375}$	$b^2c^2(c^2a^2 + a^2b^2) - (c^2a^2 - a^2b^2)^2$
σ(symmedian point)	$\sigma(X_6) = X_{76}$	b^2c^2
σ(Gergonne point)	$\sigma(X_7) = X_{330}$	$(bc - ca + ab)(bc + ca - ab)$
σ(Nagel point)	$\sigma(X_8) = X_{192}$	$ca + ab - bc$
σ(mittenpunkt)	$\sigma(X_9) = X_{6376}$	$bc(ca + ab - bc)$
σ(center of Spieker circle)	$\sigma(X_{10}) = X_{37}$	$a(b + c)$
σ(Feuerbach point)	$\sigma(X_{11}) = X_{6377}$	$a^2(ca + ab - bc)(b - c)^2$

Table 2.
$\sigma(X_i)$, for $i = 1, 2, \ldots, 11$.

name	index in ETC	first barycentric
τ(incenter)	$\tau(X_1) = X_{514}$	$b - c$
τ(centroid)	$\tau(X_2) = X_2$	1
τ(circumcenter)	$\tau(X_3) = X_{514}$	$b - c$
τ(orthocenter)	$\tau(X_4) = X_{514}$	$b - c$
τ(nine-point center)	$\tau(X_5) = X_{514}$	$b - c$
τ(symmedian point)	$\tau(X_6) = X_{1086}$	$(b - c)^2$
τ(Gergonne point)	$\tau(X_7) = X_{190}$	$(a - b)(a - c)$
τ(Nagel point)	$\tau(X_8) = X_{514}$	$b - c$
τ(mittenpunkt)	$\tau(X_9) = X_{1086}$	$(b - c)^2$
τ(center of Spieker circle)	$\tau(X_{10}) = X_{514}$	$b - c$
τ(Feuerbach point)	$\tau(X_{11}) = X_{6544}$	$(b - c)(2a - b - c)$

Table 3.
$\tau(X_i)$, for $i = 1, 2, \ldots, 11$.

Theorem 1. Suppose that ω is a symbolic substitution, given by (17):

$$p(u(a,b,c), u(b,c,a), u(c,a,b) :: ,$$

where $P = p(a,b,c) : p(b,c,a) : p(c,a,b)$ and $U = u(a,b,c) : u(b,c,a) : u(c,a,b)$ are polycenters. Then ω maps L^∞ onto L^∞.

Proof. The line L^∞ is represented as a linear combination of any two distinct polycenters on the line, such as $b - c$ :: and $a(b - c)$ ::. That is, if $X = x(a,b,c)$ :: is a polycenter on L^∞, then

$$x(a,b,c) = (b-c)(h+ak),$$

where $h = h(a,b,c)$ and $k = k(a,b,c)$ are polynomials symmetric in a,b,c, and the degree of h is 1 more than the degree of k. Then

$$\omega(X) = (u(b,c,a) - u(c,a,b)(h^* + u(a,b,c)k^*), \tag{18}$$

where

$$h^* = h(u(a,b,c),u(b,c,a),u(c,a,b)) \text{ and } k^* = k(u(a,b,c),u(b,c,a),u(c,a,b)).$$

Writing $\omega(X)$ as $x^*(a,b,c) : x^*(b,c,a) : x^*(c,a,b)$, the assumption that

$$x(a,b,c) + x(b,c,a) + x(c,a,b) = 0,$$

from (8), clearly implies that

$$x^*(a,b,c) + x^*(b,c,a) + x^*(c,a,b) = 0,$$

so that $\omega(X)$ lies on L^∞.

Next, we introduce a different kind of symbolic substitution. Call a polycenter even-powered if $p(a,b,c) = f(a^2,b^2,c^2)$ for some polynomial $f(a,b,c)$, and consider the symbolic substitution

$$SS : (a^2,b^2,c^2) \to (a,b,c).$$

For example,

$$SS(X_3) = SS(a^2(a^2 - b^2 - c^2 ::) = a(a-b-c) ::= X_9$$
$$SS(X_4) = SS(1/(a^2 - b^2 - c^2) ::= 1/(a-b-c) ::= X_7$$

Now, the Euler line is given parametrically by $sX_3 + tX_4$, where s and t are functions symmetric in a,b,c. Accordingly, the symbolic substitution SS maps the Euler line onto the line X_7X_9, or, equivalently, SS maps the line

$$[(b^2-c^2)(a^2-b^2-c^2)] \text{ onto the line } [(b-c)(a-b-c)],$$

these two lines being X_3X_4 and X_7X_9, which intersect in X_2.

A final note regarding the symbolic substitution SS is that its inverse, given by $(a,b,c) \to (a^2,b^2,c^2)$, maps polycenters to polycenters and polylines to polylines.

8. Concluding remarks

A primary focus in this chapter has been to introduce polynomial representations of triangle centers and lines (polycenters and polylines), with attention to degrees of polycenters on the Euler line, the infinity line, and the Nagel line. Sections 5–7 show the usefulness of polynomials in connection with several important *geometric* notions, including trace, combo, midpoint, isotomic conjugate, isogonal conjugate, as well as the *algebraic* notions of composition and symbolic substitution.

Much remains to be investigated; e.g., polycenters on the Brocard axis, Soddy line and others, the circumcircle and other conics, and cubics (see especially [4]). There are many geometric notions that lend themselves readily to polynomial representations, including harmonic conjugate, reflection, crosssum, crossdifference, Ceva conjugate, line conjugate, Dao conjugate, Hirst inverse, complement, anticomplement, complementary conjugate, anticomplementary conjugate, inversion in a circle, and unary operations. These notions are defined in discussed in several resources [2, 5, 8, 9, 11, 16].

Other possible extensions of the methods in this chapter include (1) classifications of families of curves, such as cubics that can be obtained via symbolic substitutions from well-established relatively simple cubics (e.g., the Neuberg, Thomson, McCay, Darboux cubics indexed as K001, K002, K003, K004 by B. Gibert [4]); (2) trigonometric polycenters, typified by the fact that $f(nA)$ :: is a polycenter for $f \in \{\text{sine, cosine, tangent, cosecant, secant, cotangent}\}$; and (3) triangle centers that are limits of sequences of polycenters (or points closely related to polycenters) but which are not themselves polycenters, such as the Hofstadter Zero Point, X_{360} in [2].

Author details

Clark Kimberling
Department of Mathematics, University of Evansville, Evansville, IN, USA

*Address all correspondence to: ck6@evansville.edu

References

[1] Davis PJ. The rise, fall, and possible transfiguration of triangle geometry: A mini-history. American Mathematical Monthly. 1995;**102**(3):204-214

[2] Encyclopedia of triangle centers (ETC). Available from: https://faculty.evansville.edu/ck6/encyclopedia/etc.html

[3] All volumes of Forum Geometricorum, 2001–2008. Available from: https://mathematicalolympiads.wordpress.com/wp-content/uploads/2012/08/forum-geometricorum-all-volumes.pdf (For example, a search of this collection quickly finds 487 appearances of the term "Euler line".)

[4] Gibert B. Cubics in the triangle plane. Available from: http://bernard-gibert.fr/

[5] Glossary (Encyclopedia of Triangle Centers). Available from: https://faculty.evansville.edu/ck6/encyclopedia/glossary.html

[6] Rabinowitz S. Inequalities derived from distances between triangle centers. International Journal of Computer Discovered Mathematics. 2022;**7**:181-194. Available from: http://www.journal-1.eu/index.htm

[7] Kimberling C. Symbolic substitutions in the transfigured plane of a triangle. Aequationes Mathematicae. 2007;**73**:156-171

[8] Lozada CE. Alphabetical index to terms in ETC. Available from: https://faculty.evansville.edu/ck6/encyclopedia/Alphabetical_Index.html

[9] MathWorld. Available from: https://mathworld.wolfram.com

[10] Moses PJC. Central Lines. Available from: https://faculty.evansville.edu/ck6/encyclopedia/CentralLines.html

[11] Moses PJC, Kimberling C. Unary operations on homogeneous coordinates in the plane of a triangle. Geometry. 2024;**1**:1-14. Available from: https://www.mdpi.com/3042-402X/1/1/2

[12] Odehnal B. Poristic loci of triangle Centers. Journal of Geometry and Graphics. 2017;**15**:45-67

[13] The on-line encyclopedia of integer sequences. Available from: http://oeis.org

[14] Reznik D, Garcia R, Koiller J. Can the elliptic billiard still surprise us? The Mathematical Intelligencer. 2019;**42**(1):6-17. Available from: https://link.springer.com/article/10.1007/s00283-019-09951-2

[15] Sandifer CE. 19th century triangle geometry, in How Euler Did It, The Mathematical Association of America. 2007;3:19-27. Available from: http://eulerarchive.maa.org/hedi/HEDI-2006-05.pdf

[16] Yiu P. (2001, 2013) Introduction to the Geometry of the Triangle. Available from: https://web.archive.org/web/20180422091419id_/http://math.fau.edu/Yiu/YIUIntroductionToTriangleGeometry130411.pdf

Chapter 2

Splines for Fitting Curves and Surfaces of Nonlinear Functions

Constantin Volosencu

Abstract

The chapter is placed in the broad context of geometric fitting of parametric curves and surfaces of nonlinear functions. The purpose of the study is to compare some methods of fitting in the case of the example of some nonlinear functions, such as the exponential function and the *sinc* multivariable function. The methods used are spline, interpolation, smoothing and thin plate splines and respectively polynomial and piecewise cubic interpolation. The main results are the graphs of the fitting curves and the residuals for these methods as a basis for comparison. MATLAB tools were used for calculations.

Keywords: Cubic spline interpolation, least-squares approximation, smoothing spline, thin-plate splines, curve fitting, sinc function, polynomial fitting, piecewise cubic interpolant

1. Introduction

The interpolation of functions is solved in practice through a multitude of mathematical solutions. The use of spline functions ensures good precision and simple and easy-to-use equations. The notion of spline implies the use of functions composed of polynomials defined. The main applications of the spline are its use in the interpolation of functions, curve and surface fitting, computer-aided design, and computer graphics. The functions on which the interpolation operation is performed can be mono-variable or multivariable, and they are generally strongly nonlinear. These are interpolated with spline functions, which are finite dimensional [1–5]. This study is placed in a broad context of broad context of geometric fitting of parametric curves and surfaces of nonlinear functions. The purpose of the work is to compare some fitting methods on the case of this mathematic nonlinear function with significance in curve and surface fitting. Fitted curves may be used for data visualization, to infer values of function where no data are available and to summarize the relationships among the variables. Extrapolations may be done by using the fitted surface beyond the range of the observed data, with a degree of uncertainty, which is reflected by the method used to construct the surface. The controversial and diverging hypotheses consist of the quality of the approximation and the characteristics of the methods used.

Some recent studies using spline functions are briefly reviewed below. Recent research in polynomials and approximation usable in curve fitting are presented in Ref. [6]. A recent study on generalized quantum polynomial is presented in Ref. [7]. The book [8] provides some recent studies on splines and their applications. The chapters published in this books include information in computer-aided design, disease detection, computer vision, curve fitting, regression curves, power systems, social media, and biology. The studies presented in the book are supported by computer programs. In the paper [9], it is shown that spline models are effective in engineering and biomedical sciences, in the case of processing complex signals covered by noises. Thus, a spatial representation of the states of the variant analysis models is introduced, using estimation algorithms based on the Kalman filter, making numerical calculation more efficient. In the paper [10], hyperbolic spin interpolation is used to develop a numerical model to obtain approximate solutions of the generalized Burgers-Fisher equation. Several numerous examples are presented whose results are compared with those in the literature. In the paper [11] presents a study of application pf spline in the field of wheeled mobile robots to plane smooth paths for precise vehicle movements, to obtain good trajectories. The concept of third-order geometric continuity for Cartesian curves and paths is used. Several path generation examples are presented. In the paper [12], an application of cubic b-spline for total electron content mapping is presented, in the field of the earth-space communication systems, for transmission through the ionosphere and skywave systems with ionospheric refraction. Spline functions can be used to approximate curves and surfaces that describe the behavior of fuzzy systems, as in Refs. [13, 14]. Also, spline functions can be used in fitting the state curves of distributed parameter systems [15].

For curve fitting, an exponential curve is chosen from mechanical engineering applications and four methods were compared: cubic spline interpolation, least-squares approximation, smoothing spline and thin-plate splines. For surface fitting, the work makes a comparison of the results obtain with ANFIS, polynomial interpolation, and piecewise cubic interpolant methods. The chapter is kept also to scientists outside the field of research. The paper presents an example for the process of constructing a surface of the *sinc* nonlinear multivariable function using two methods: polynomial and piecewise cubic interpolant. The examples present the fit to a series of data points. MATLAB tools were used for calculations [16].

2. Curve fitting

2.1 Methods

We are choosing to fit of an exponential curve, form often encountered in the design of mechanical parts which can be, for example, in practice, the shape of ultrasonic waveguides:

$$y = e^{-x}, x \in [0, 1] \tag{1}$$

Some theoretical aspects of the methods used are presented below. The bibliography for these methods can be consulted in Refs. [13, 14].

2.1.1 Splines

The spline functions may be defined using a set:

$$S := \Pi_{b,k}^{m} \tag{2}$$

of piecewise polynomials of order k with breaks $b_1 < \ldots < b_{p+1}$, that have no jump in any lower-order derivative than mth derivative. This set is a linear one. And of course it contents a basis, a sequence $f_1,\ldots,f_n$ and every f in S can be written uniquely in the form:

$$f(x) = \sum_{i=1}^{n} a_i f_i(x) \quad (3)$$

with the coefficients a_i. and n is the dimension of the linear space S. The coefficients a_i. are the coordinates of f with respect to this basis. The space S has a basis consisting of B-splines.

For cubic spline the interpolant function is a third-degree polynomial defined by

$$p_i(x) = a_i(x - x_i)^3 + b_i(x - x_i)^2 + c_i(x - x_i) + d \quad (4)$$

where for p_i the variable x_i takes values on intervals $x_{i-1} \le x < x_i$, $i = 1, \ldots, n$-1.
A piecewise-polynomial function is a function built from polynomials.

2.1.2 Interpolation

It is a procedure to build a function f that matches a set of data values (x_i, y_i): $f(x_i) = y_i$, for all i. The function f is the interpolant, developed as a unique function of the form (3).

Polynomial interpolation may be chosen because for n data points (xi, yi) there is one polynomial of order n–1 that matches these data:

$$f_{i(x)=\prod_{i \neq j}(x-x_j)} \quad (5)$$

In spline interpolation, one chooses the f_j to be the n consecutive B-splines, using Schoenberg-Whitney theorem.

2.1.3 Least-squares approximation

In least-squares approximation the linear system:

$$y_i = f(x_i) = \sum_j f_j(x_i), i = 1, \ldots, n \quad (6)$$

is solved based on the least-squares method. The coefficients, as weighting, of the approximation function are determined so as to minimize the error measure:

$$E(f) = \sum_i w_i \left(y_i - f(x_i)\right)^2 \quad (7)$$

The data may be matched only approximately.

2.1.4 Smoothing

In spline smoothing, one also tries to make such an error measure small, but tries, at the same time, to keep the following roughness measure small. For data values y_i at sites c_i in the plane, one uses instead the error measure and roughness measure

$$E(f) = \sum_i \left(y_i - f(c_i)\right)^2 \tag{8}$$

with

$$F(D^2 f) = \int \left(|D_{11}f|^2 + 2|D_{12}f|^2 + |D_{22}f|^2\right) \tag{9}$$

where the minimizer of the sum is a thin-plate spline.

2.1.5 Thin-plate splines

A bivariate thin-plate spline has the expression:

$$f(x) = \sum_{i=1}^{n-3} \varphi\left(|x - c_i|^2\right) a_i + x(1) a_{n-2} + x(2) a_{n-1} + a_n \tag{10}$$

where φ is a univariate function, and c_i are the centers.

2.2 Results

To make calculation for this example, the spline tool designer app from MATLAB 2018 was used. This app let the user to make experiments with some spline approximation methods, as it follows: fit cubic spline interpolants to curves and surfaces, fit smoothing splines and shape-preserving cubic spline interpolants to curves, and fit thin-plate splines to surfaces. The computations may be done using specific splines functions to fit, as: B-form, tensor-product, and rational and thin-plate splines. The breaks may be specified. The knot placement may be optimized. Differentiation and integration may be used for spline manipulation.

The definition domain of the function was divided into $n = 10$ parts and 11 points were chosen for trajectory approximation of the function (1). The results obtained using four methods: Cubic spline interpolation, least square approximation, smoothing spline, and spline approximation are presented in **Figures 1–4**.

The first graph shows the dataset chosen for approximation and the shape of the fitting curve, and the second graphic shows the curve of the variation of the fitting errors.

3. Surface fitting

3.1 Preliminaries

The fitting method as spline has a good capability of fitting a nonlinear function because due to the fact that it is a combination of many polynomials on many parts of the function. It is possible to use surface fitting for modeling, approximating or estimating a nonlinear function of several variables, and the obtained result will have a small error depending on the fitting method and multiple conditions of the fitting process, such as the amount of training data used and depending on the available computing power and others.

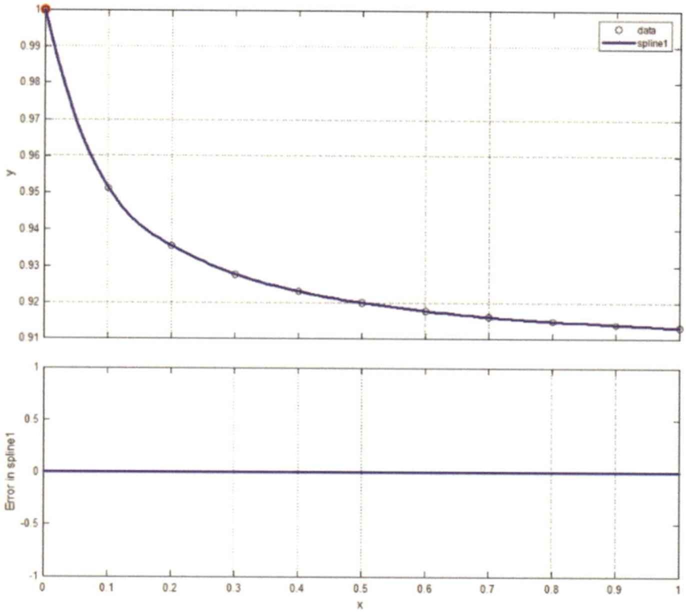

Figure 1.
Cubic spline interpolation.

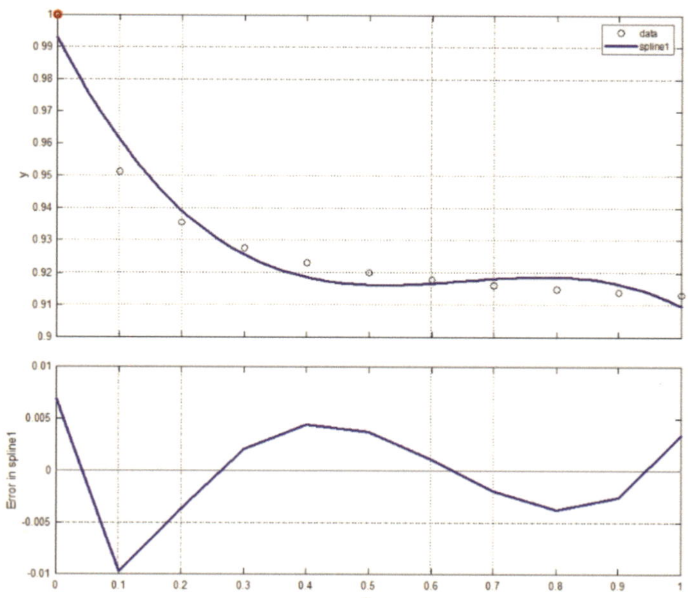

Figure 2.
Least square approximation.

We are choosing a nonlinear function with two variables, the *sinc* function:

$$z = \frac{10\sin(r)}{r}, r = \sqrt{x^2 + y^2}, x \in [-10, 10], y \in [-10, 10] \tag{11}$$

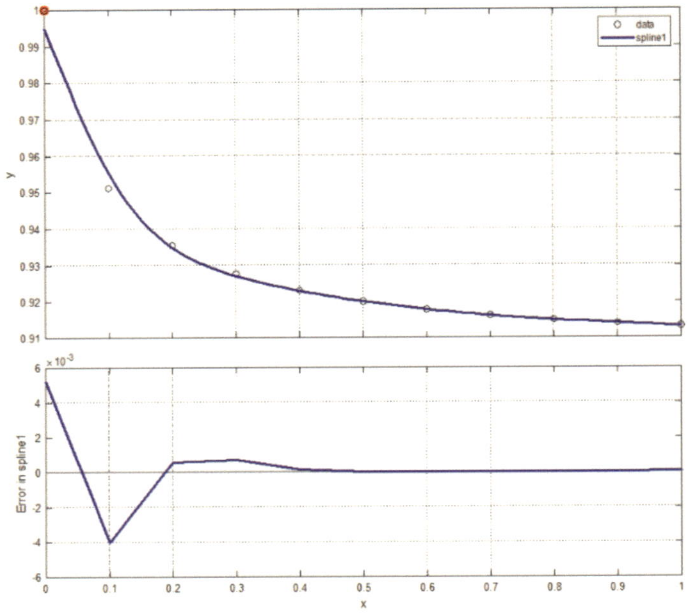

Figure 3.
Smoothing spline.

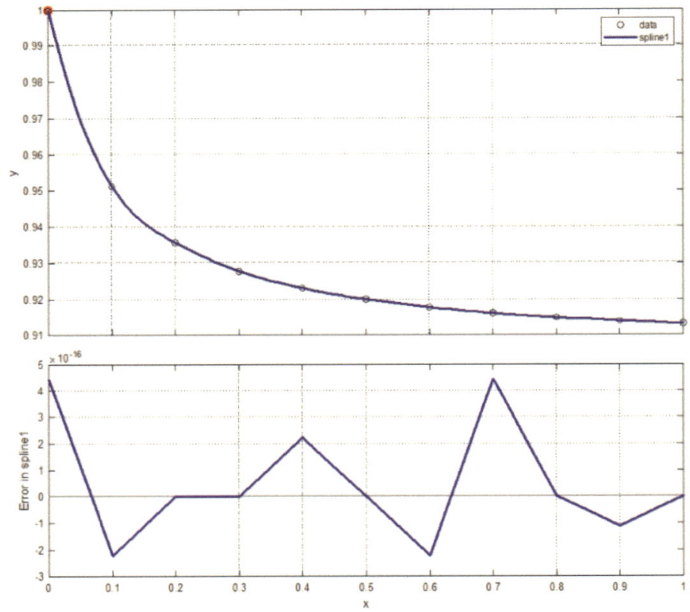

Figure 4.
Spline approximation.

Figure 5 shows the function sinc used to generate the data.

The chosen example of fitting a nonlinear multivariable function was used in Ref. [17] with a neuro-adaptive learning. **Figures 6** and **7** show the results obtained as a basis for comparison.

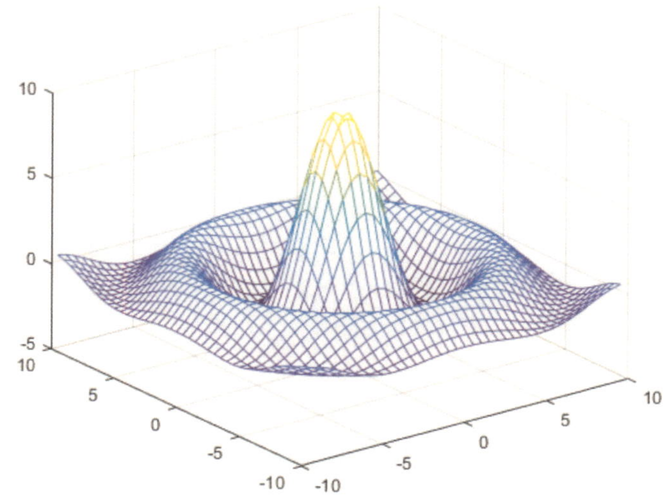

Figure 5.
The graph of the sinc *function.*

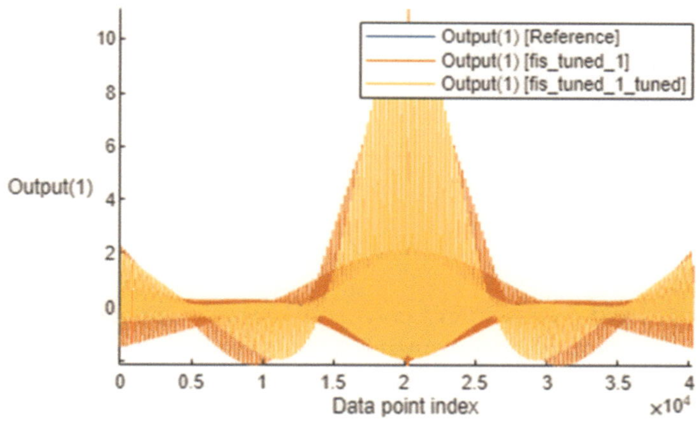

Figure 6.
The graph of the sinc *function and its approximation with ANFIS.*

In practice, better results can be obtained in matching the surface with interpolation methods, as will be seen below.

3.2 Methods

3.2.1 Polynomial fitting

The surface is fitted using a polynomial function. The values of the coefficients were determined by fitting the polynomial to the training data, by minimizing an error function that measures the misfit between the function, for any given value of variables, and the training set data points. The error function is the sum of the squares of the errors between the predictions for each data point and the corresponding target values. The curve fitting problem is solved by choosing the value of coefficients for which error is as small as possible. Since the error function is a quadratic function of

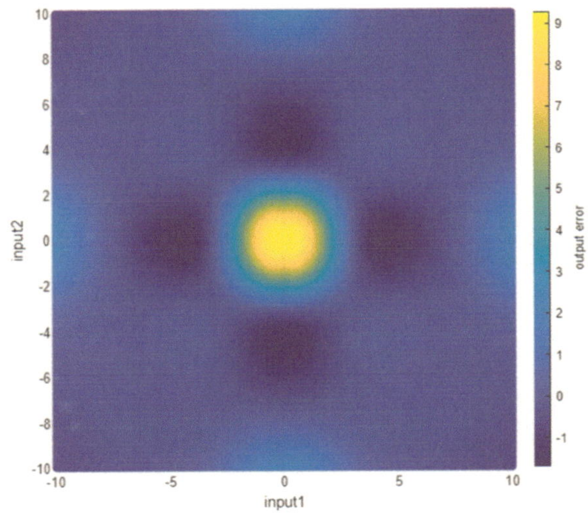

Figure 7.
The graph of approximation errors for ANFIS.

the coefficient, its derivatives with respect to the coefficients will be linear in the elements of coefficients, and so the minimization of the error function has a unique solution.

A training set of nxn samples of function $z = f(x,y)$:

$$x = (x_1, x_2, \dots x_n), y = (y_1, y_2, \dots y_n), z = \begin{pmatrix} z_{1,1} & \cdots & z_{1,n} \\ \vdots & \ddots & \vdots \\ z_{n,1} & \cdots & z_{n,n} \end{pmatrix} \quad (12)$$

was generated and used to train a polynomial function, to make a prediction according to a criterion. The polynomial is a nonlinear function in x and y but a linear one in its parameters grouped in a vector w, with the properties of a linear model. A higher order of the polynomial is not a better one. A fit is obtained by minimizing an error function $E(w)$, sum of squares of the errors between the predictions for each data point $z(x, y, w)$ and target value:

$$E(w) = \frac{1}{2}(z(x, y, w) - t)^2 \quad (13)$$

The coefficients solution w of error is calculated. The error function E is quadratic in coefficients w. Its derivatives are linear in w and they have a unique solution w^*.

$$\frac{\partial E(w)}{\partial w_i} = 0 \quad (14)$$

The resulting polynomial is $z(x, y, w^*)$. Least square is case of maximum likelihood. If the size of data increases the model overfitting is less severe. The data should be adapted to parameter model and the model complexity is recommended to be chosen according to problem complexity.

The order of the polynomial was chosen from 1 to 3. The third-order polynomial seems to give a good fit to the function of the example. Magnitude of coefficients increases as order of polynomial increases. Large positive and negative values resulted so that the corresponding polynomial function matches each of the data points exactly, but between data points the function exhibits the large oscillations. This is over-fitting. Increasing the size of the data set reduces the over-fitting problem.

To make calculation for this example, the curve fitting toolbox *cftool* from MATLAB 2018 was used. This toolbox let the user to design, train, and test different fitting methods, using fitting dataset of the multivariable functions.

3.2.2 Cubic interpolant

In this method also the same data samples were used. An error bound is used. High-order polynomials have oscillations in their behavior. The piecewise could be smoothness. The changing in data could alter the polynomial interpolant. The chosen interval of data set [a, b] is divided into subintervals:

$$[a, t_1, t_2, \dots, t_{m-1}, b] \tag{15}$$

With $a = t_0$ and $b = t_m$, where t_i are breakpoints or knots.

The interpolants are constructed:

$$v(x,y) = p_{i,j}\left(x_i, y_j\right), t_i \le x_i \le t_{i+1}, t_j \le y_j \le t_{j+1} \tag{16}$$

Where each $p_{i,j}(x_i, y_j)$ is a polynomial of low degree, piecewise cubic with $n = 3$.

$v(x, y)$ must satisfy the interpolation conditions, with a global smoothness property.

Local polynomial interpolants are constructed as simple function using local polynomial interpolation on each subinterval $[t_i, t_{i+1}]$. Polynomial interpolation assures derived error bounds. Locally data influences the interpolant. Of course, there are a lot of limitations on quality fitting. Errors analysis may be done for piecewise approximation.

3.3 Results

A training dataset of $z(x_i, y_j)$, with n = 40 points for x and y variables and 1600 points for z was chosen.

3.3.1 Polynomial

The resulting polynomial is:

$$z = f(x,y) = p_{00} + p_{10}x + p_{01}y + p_{20}x^2 + p_{11}xy + p_{02}y^2 + p_{30}x^3 + p_{21}x^2y + p_{12}xy^2 + p_{03}y^3 \tag{17}$$

where the coefficients (with 95% confidence bounds) are: p_{00} = 0.6808 (0.5294, 0.8322), p_{10} = 0.005132 (−0.03311, 0.04337), p_{01} = 0.005132 (−0.03311, 0.04337), p_{20} = −0.00781 (−0.01055, −0.00507), p_{11} = −8.268e-05 (−0.002524, 0.002359), p_{02} = −0.00781 (−0.01055, −0.00507), p_{30} = 1.714e-05 (−0.0005181, 0.0005524), p_{21} = −0.0001763 (−0.0006453, 0.0002926), p_{12} = −0.0001763 (−0.0006453, 0.0002926), p_{03} = 1.714e-05 (−0.0005181, 0.0005524).

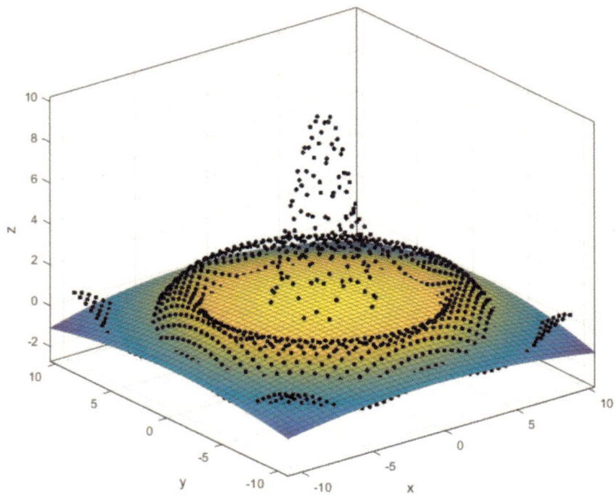

Figure 8.
The graphs of the fit results for polynomial method.

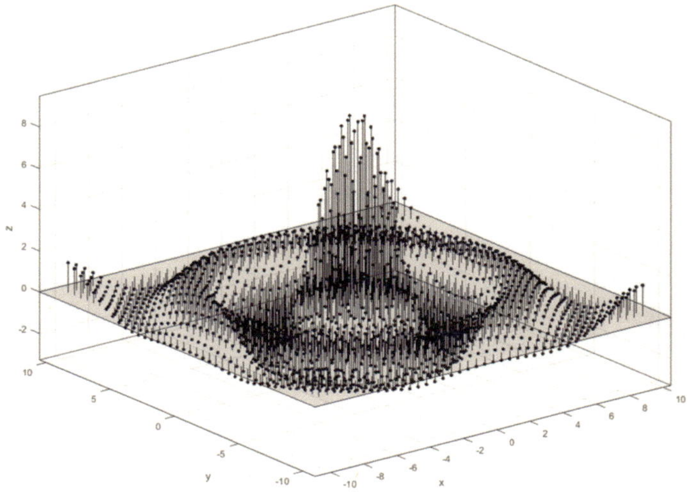

Figure 9.
The graph of the residuals for polynomial method.

The goodness of fit is represented by SSE: 4290, R-square: 0.03944, and adjusted R-square: 0.034 (**Figures 8** and **9**).

3.3.2 Cubic interpolant

In the piecewise cubic interpolant, the fitting function is z = f(x,y) is a piecewise cubic surface computed from coefficients, where x is normalized by mean − 0.2502 and std. 5.775 and y is normalized by mean − 0.2502 and std. 5.775. The goodness of fit is represented by SSE: 0, R-square: 1, DFE: 0 and coeff: 1599 (**Figures 10** and **11**).

Results with the same quality of surface fitting may also be obtained with the biharmonic interpolant method.

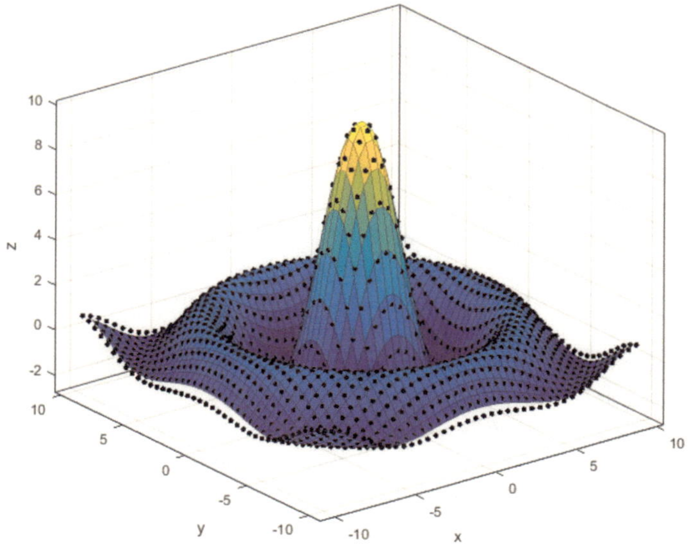

Figure 10.
The graph of the fit results for cubic method.

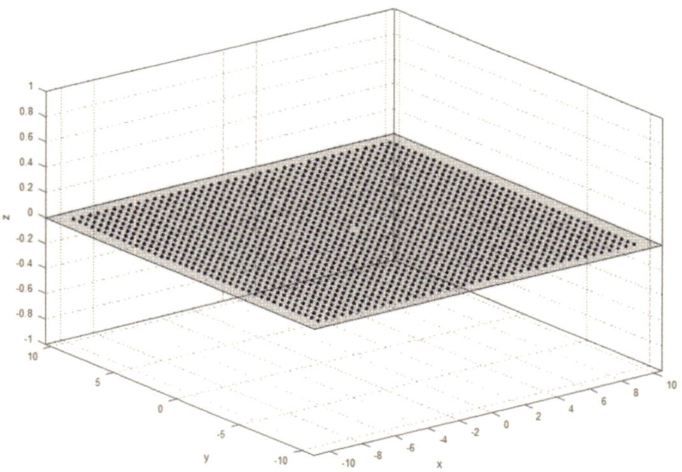

Figure 11.
The graph of the residuals for cubic interpolant method.

4. Conclusion

The chapter presents some applications of spline methods and numerical examples of applying the theory of spline functions in fitting the curves and surfaces of the graphs of nonlinear functions. This examples may be taken in consideration as a base for computer-aided design. Four curves fitting the following methods: cubic spline interpolation, least square approximation, smoothing spline, and spline approximation were tested on an example of curve fitting of an exponential function, the form often encountered in the design of mechanical parts. The results of the calculation consist in the graphs of dataset chosen for approximation and the shapes of the fitting curves, and the graphs of the variation of the fitting errors. Cubic spline interpolation and spline interpolation have the minimum errors, throughout the range of values. Smoothing spline has greater errors for greater values of the exponential function. And least square approximation has the greatest errors throughout the range of values. The approximation errors vary a lot on the curve of the nonlinear function, being on certain short portions small, acceptable, but on other portions quite large, assuming the choice of another method. The duration of calculation is very small. After these tests, some of these methods could be chosen having to be reduced through other attempts. It is likely that by doing other tests for other functions certain methods of the above may be chosen.

Also, the chapter presents an example of the curve fitting of a nonlinear function based on polynomial interpolation and cubic interpolant. The chosen function is the *sinc* function, a nonlinear function of two variables. An ANFIS approximation was used for comparison. To make calculation for this example, the curve fitting toolbox cftool from MATLAB 2018 was used. From this toolbox the two fitting methods were chosen. The fitting procedures were designed, trained, and tested. A set of fitting data was chosen. Training results are presented: the graphs of the fit results and of the residuals for polynomial method and for the cubic method. The polynomial structure is a 3rd grade one, with 2 variables and 9 coefficients. The method with cubic interpolant has better results, with a good curve fitting and small residuals. The graph of the residuals for cubic piecewise interpolant method demonstrates this fact.

Author details

Constantin Volosencu
"Politehnica" University, Timisoara, Romania

*Address all correspondence to: constantin.volosencu@aut.upt.ro

References

[1] de Soor. CA Practical Guide to Splines. New York, NY: Springer; 2001. p. 348

[2] Arlinghaus SL. PHB Practical Handbook of Curve Fitting. Boca Raton: CRC Press; 1994

[3] Birkhoff G, de Boor C. Piecewise polynomial interpolation and approximation. In: Approximation of Functions. Amsterdam: Elsevier Publishing Company; 1964. pp. 164-190

[4] Sung-Joon A. Geometric fitting of parametric curves and surfaces. Journal of Information Processing Systems. 2008;**4**:153-158

[5] Carlson RE, Hall CA. On piecewise polynomial interpolation in rectangular polygons. Journal of Approximation Theory. 1971;**4**:31-53

[6] Özger F, editor. Recent Research in Polynomials. London, UK: Intechopen; 2023. 194 p. DOI: 10.5772/intechopen.1001517

[7] Silindir B, Yantir A. Generalized quantum polynomials. In: Özger F, editor. Recent Research in Polynomials. London: IntechOpen; 2023. DOI: 10.5772/intechopen.1000861

[8] Truong YKN, Sarfraz M, editors. Topics in Splines and Applications. London: InTech; 2018. DOI: 10.5772/intechopen.68737

[9] Parker J, Zhang Y, Lafleur JB, Sun X. Computationally efficient Kalman filter approaches for fitting smoothing splines [Internet]. In: Kim YV, editor. Kalman Filter—Engineering Applications. London, UK: IntechOpen; 2023. DOI: 10.5772/intechopen.106713

[10] Jeyar M, Lamnii A, Yassir Nour M, Oumellal F, Zidna A. An algebraic hyperbolic spline quasi-interpolation scheme for solving burgers-fisher equations. In: Volosencu C, Cheon Seoung Ryoo CS, editors. Simulation Modeling. London, UK: Intechopen; 2022. DOI: 10.5772/intechopen.99033

[11] Piazzi A, Lo Bianco CG, Romano M. Smooth path generation for wheeled Mobile robots using eta;3-splines. In: Casolo F, editor. Motion Control. London UK: InTech; 2010. DOI: 10.5772/intechopen.6960

[12] Grynyshyna-Poliuga O, Stanislawska I, Swiatek A. Regional ionosphere mapping with kriging and B-spline methods. In: Notarpietro R, Dovis F, De Franceschi G, Marcio Aquino M, editors. Mitigation of Ionospheric Threats to GNSS: An Appraisal of the Scientific and Technological Outputs of the TRANSMIT Project. London, UK: InTech; 2014. DOI: 10.5772/intechopen.58776

[13] Zakaria R, Wahab AF, Gobithaasan RU. Fuzzy B-Spline Surface Modeling. Journal of Applied Mathematics, Hindawi Publishing Corporation. 2014. p. 8. Article ID: 285045. DOI: 10.1155/2014/285045

[14] Volosencu C. Stabilization of fuzzy control systems. WSEAS Transactions on Systems and Control. 2008;**10**(3): 879-896

[15] Volosencu C. Identification of distributed parameter systems, based on sensor networks and artificial intelligence. WSEAS Transactions on Systems. 2008;**6**(7):785-801

[16] Curve Fitting Toolbox™ User's Guide. Natick, MA: The MathWorks,

Inc; 2019. Available from: https://www.mathworks.com/help/pdf_doc/curvefit/curvefit.pdf

[17] Volosencu C. Introductory chapter: ANFIS for modeling multivariable nonlinear functions. In: Volosencu C, editor. Adaptive Neuro-Fuzzy Inference System as a Universal Estimator. London: Intechopen; 2024. DOI: 10.5772/intechopen.1004337

Chapter 3

Iterative Techniques for Nonlinear Equations: Addressing Multiple Roots with Unknown Multiplicity

Saima Akram, Moin-Ud-Din Junjua, Faiza Akram and Laiba Khalid

Abstract

Over time, the need for precise and effective solutions to complex real-world phenomena has grown increasingly important. Dealing with multiple roots can be quite challenging, especially when the multiplicity is unknown. The sense of urgency arises from the rapid progress made in different areas of science and engineering. In this chapter, we discuss various multipoint iterative techniques based on various approaches such as Newton-Househölder approach, weight function approach, and parameter approach to develop optimal eighth-order methods for multiple roots with unknown multiplicity. Their convergence is analyzed and the higher convergence order leads to a faster attainment of the root with a high level of precision. The numerical results are calculated and compared w.r.t computational order of convergence, difference between successive iterations, and asymptotic error constant for these iterative schemes. This comparison effectively validates the stability of theoretical predictions and observes favorable convergence behavior exhibited by the newly proposed schemes. Various test functions are utilized to check the effectiveness of the newly developed iterative schemes, especially for model of osteoporosis in Chinese women, beam designing and root clustering problem. The numerical results depict that new iterative functions are highly competitive when compared to established optimal iterative approaches. Hence, they can be good addition to the literature.

Keywords: iterative methods, nonlinear polynomials, order of convergence, efficiency index, multiplicity

1. Introduction

Numerical analysis is a branch of mathematics dedicated to designing and analyzing algorithms that provide numerical solutions to complex mathematical problems. These problems frequently emerge from real-world applications across various disciplines such as science, engineering, and finance. To address these challenges, numerical methods employ a range of techniques, including interpolation, differentiation, integration, and the solving of both linear and nonlinear equations.

These methods are crucial for tackling real-world issues such as continuous stirred tank reactors (CSTR), beam design problem, population growth model, and fractional conversion in chemical reactors, where exact solutions are often impractical or impossible due to the complexity of the equations involved. By using iterative processes and approximation techniques, numerical analysis provides powerful tools to navigate these complexities.

Key aspects of numerical analysis include the following:

Accuracy: This refers to how closely the numerical solution approximates the exact solution.

Stability: This pertains to how an algorithm responds to small changes in input or intermediate calculations.

Efficiency: This involves the computational resources needed to achieve the desired level of accuracy.

In the following sections, we will introduce some fundamental definitions essential for understanding numerical analysis.

1.1 Preliminaries

1.1.1 Algebraic equation

An algebraic equation is a mathematical statement asserting that two algebraic expressions are equivalent [1]. It typically includes variables, constants, and arithmetic operations, and is characterized by the equal symbol (=) between two algebraic expressions. Examples of algebraic equations are:

$$k^2 + 4k - 5 = 0, \tag{1}$$

$$2k + 5 = 11. \tag{2}$$

1.1.2 Polynomial

A polynomial $P(k)$ in the variable k is expressed in the form [1]:

$$P(k) = a_t k^t + a_{t-1} k^{t-1} + a_{t-2} k^{t-2} + \ldots + a_1 k + a_0, a_t \neq 0, \tag{3}$$

where $a_t, a_{t-1}, \ldots, a_1, a_0$ are constants (coefficients), t is a non-negative integer (the degree of the polynomial), and k is the variable.

1.1.3 Nonlinear equation

A nonlinear equation is defined as an equation of the form $f(k) = 0$ where the highest degree of k is 2 or greater [1]. Transcendental equations are also categorized as nonlinear equations. Examples include:

$$k^2 + k + 2 = 25, \tag{4}$$

$$y - \sin(k) + e^k = 0. \tag{5}$$

1.1.4 Iterative method

Iterative methods are mathematical techniques that use repeated calculations to find solutions to problems [1]. The iterative method for successively approximating the root of $f(k) = 0$ is described as:

$$k_{t+1} = \phi(k_t), t \geq 0, \tag{6}$$

where $\phi(k_t)$ denotes the iteration function. The approximations of the $(t)^{th}$ and $(t+1)^{th}$ iterations to the root α are k_t and k_{t+1}, respectively. The process typically begins with an initial guess k_0 and refines through successive iterations until the desired accuracy or convergence is achieved.

1.1.5 Multiplicity

A root of the equation $f(k) = 0$ is said to have a multiplicity m if the function f can be expressed as [2]:

$$f(k) = (k-\alpha)^m h(k), \tag{7}$$

where $h(k)$ is a function such that $\lim_{k\to\alpha} h(\alpha) \neq 0$. For example:

$$(k+3)^2(k-2) = 0, \tag{8}$$

has a root of multiplicity 2 at $k = -3$ and a root of multiplicity 1 at $k = 2$. A simple root has a multiplicity of one.

1.1.6 Efficiency index

The efficiency index of an iterative scheme measures how effectively numerical methods solve nonlinear equations [3]. This index considers both the order of convergence and the computational cost, which is quantified by the number of function evaluations. The efficiency index (EI) of an iterative method is defined as:

$$EI = r^{\frac{1}{t}}, \tag{9}$$

where r represents the order of convergence of the iterative method and t denotes the number of function evaluations and its derivative per iteration.

1.1.7 Optimal convergence order

The optimal convergence order of a multipoint iterative method is achieved when $t+1$ derivative or function evaluations per iteration yield a maximum convergence order of 2^t [4].

2. Literature review

A significant challenge in numerical analysis is the approximation of roots for solving nonlinear equations, especially when exact solutions are difficult to find. Abel's 1824 statement highlighted the complexity of computing exact solutions for polynomials of degree five or higher, thus necessitating the development of approximate methods. Iterative methods are employed to approximate solutions by starting with an initial guess and refining it through repeated iterations. These methods are particularly beneficial for nonlinear equations with multiple roots, as they can converge to a solution even when analytical solutions are impractical. For instance, Newton's method (10) is a well-known technique for finding the roots of a single nonlinear equation $f(k) = 0$, and is described as follows:

$$k_{t+1} = k_t - \frac{f(k_t)}{f'(k_t)}, t \geq 0, \tag{10}$$

where f is a real-valued function. This method employs a one-point iterative approach, demonstrating quadratic convergence in the vicinity of the root α of f, provided the initial guess k_0 sufficiently close to α.

When dealing with nonlinear equations that have multiple roots, iterative methods, such as Newton's method (10), face significant challenges. Newton's method in particular struggles with multiple roots because the derivative $f'(k)$ at these roots can be very small, leading to slow or even divergent convergence. Furthermore, the method necessitates a good initial guess, which is especially critical near multiple roots.

The primary issues encountered with iterative methods in the context of multiple roots include:

Convergence difficulties: When roots are closely spaced, iterative methods may struggle to converge to a specific root, making it difficult to determine which root to approach.

Initial guess sensitivity: The accuracy of the initial guess plays a pivotal role. An inaccurate initial guess can result in convergence to an unintended root or significantly reduce the efficiency of the process.

High multiplicity: Roots with high multiplicity can slow convergence or cause failure due to their small derivatives. Additionally, numerical errors can become significant with multiple roots, affecting stability.

Numerical errors: Multiple roots can exacerbate numerical errors, affecting the stability of the solution. These errors may cause the iterative method to explore various paths to different roots, complicating the verification of all possible solutions.

Traditional methods, such as Newton's method, often perform poorly with multiple roots because they typically assume that each root is simple and occurs only once. This assumption is not valid for multiple roots, leading to ineffective or inefficient solutions. Consequently, finding and understanding multiple roots is a complex task, particularly when the multiplicity of the root is unknown.

The earliest modification of Newton's method for approximating multiple roots, Rallís method [5], was introduced by Schröder [6]. This modification is expressed by the following formula:

$$k_{t+1} = k_t - m \cdot \frac{f(k_t)}{f'(k_t)}, \tag{11}$$

where m denotes the multiplicity of the root and must be known a prior. This method converges quadratically. Although significant advances have been made in methods for finding simple roots, fewer methods are available for handling multiple roots of nonlinear equations. In recent years, following the modification of Newton's method for multiple roots, various schemes have been developed to accurately locate multiple roots of nonlinear equations when the multiplicity is known. Extensive research has led to the creation of numerous methods for this purpose, and a range of approaches is detailed in the literature [7–19]. In the year 2021, Akram et al. [20]

developed an eighth-order method for finding multiple roots of nonlinear equations with known multiplicity.

2.1 Optimal eighth-order iterative scheme for finding multiple roots with known multiplicity

In 2021, Akram et al. [20] introduced an iterative method designed to efficiently approximate multiple roots of nonlinear equations. Consider α as the multiple root with a known multiplicity $(m \geq 1)$ of the function $f : \mathbb{C} \to \mathbb{C}$ within a neighborhood of α. The proposed method features a simple structure, employing a weight function approach in each step of the iterative scheme:

$$\begin{aligned} y_t &= k_t - G(v_t), t \geq 0, \\ z_t &= y_t - H(p_t) \cdot G(v_t), \\ k_{t+1} &= z_t - H(p_t) \cdot D(p_t, s_t, w_t) \cdot G(v_t), \end{aligned} \tag{12}$$

where $v_t = \frac{f(k_t)}{f'(k_t)}, p_t = \left[\frac{f(y_t)}{f(k_t)}\right]^{\frac{1}{m}}, s_t = \left[\frac{f(z_t)}{f(y_t)}\right]^{\frac{1}{m}}, w_t = \left[\frac{f(z_t)}{f(k_t)}\right]^{\frac{1}{m}}$. Here, $G : \mathbb{C} \to \mathbb{C}$ was a differentiable weight function in the neighborhood of (0).

The univariate weight function $H : \mathbb{C} \to \mathbb{C}$ and the trivariate weight function $D : \mathbb{C}^3 \to \mathbb{C}$ were analytic in the neighborhood of (0) and $(0, 0, 0)$, respectively.

The presented scheme (12) achieved the optimal eighth order of convergence.

Theorem 1.1 Let α be a multiple root with multiplicity $(m \geq 1)$ of the involved function f and $f : \mathbb{C} \to \mathbb{C}$ is analytic in the region enclosing a multiple zero α of f. Moreover, assuming that $D : \mathbb{C}^3 \to \mathbb{C}, G : \mathbb{C} \to \mathbb{C}$, and $H : \mathbb{C} \to \mathbb{C}$ are analytic in the neighborhood of their respective origins. Given an initial guess k_0 that is sufficiently close to α for f, the iterative method defined by (12) achieves an optimal eighth order of convergence under the following conditions:

$A_0 = 0; A_1 = k; A_2 = 0; B_0 = 0; B_1 = 1; B_2 = 4; B_3 = 18; D_{000} = 0; D_{100} = 0;$ - $D_{010} = 1; D_{200} = 0;\ D_{001} = -1; D_{110} = 1; D_{020} = 2; D_{101} = 1; D_{011} = 2,$

and error equation is given by:

$$\begin{aligned} e_{t+1} = &\frac{1}{24m^7}\left[C_1\left((3+m)C_1^2 - 2mC_2\right)\left(\left(-163 + 7m^2\right)C_1^4 - 24m^2C_1^2C_2 - 12m^2C_2^2 + 12m^2C_1C_3\right)e_t^8\right] \\ &+ O\left(e_t^9\right), \end{aligned} \tag{13}$$

where $e_t = k_t - \alpha$ and $C_t = \frac{m!}{(m+t)!}\frac{f^{(m+t)}(\alpha)}{f^{(m)}(\alpha)}$, for $t \in \mathbb{N}$.

2.1.1 Special case of weight function $(SF-2)$

We consider a particular case of the approach (12) and is given as follows:

$$\begin{aligned} y_t &= k_t - (m \cdot v_t), t \geq 0, \\ z_t &= y_t - \left(p_t + 2p_t^2\right) \times (m \cdot v_t), \\ k_{t+1} &= z_t - \left(p_t + 2p_t^2\right) \times (m \cdot v_t) \times \left(s_t \cdot \left(1 + s_t + 2s_tp_t + p_t^2\right) - 6w_tp_t^2\right), \end{aligned} \tag{14}$$

where $v_t = \frac{f(k_t)}{f'(k_t)}, s_t = \sqrt[m]{\frac{f(z_t)}{f(y_t)}}, p_t = \sqrt[m]{\frac{f(y_t)}{f(k_t)}}$ and $w_t = \sqrt[m]{\frac{f(z_t)}{f(k_t)}}$. We named it $SF-2$.

There is a scarcity of literature on iterative schemes for finding higher-order multiple roots, especially when the multiplicity of these roots is unknown. The primary challenge lies in the complexity and time-consuming nature of developing iterative methods for handling repeated roots. Most existing techniques for multiple roots are extensions of the modified Newton's method, often involving complex structures.

However, much less work has been done on developing methods using Traub's conceptual approach. Traub [21] addressed the challenge of approximating a multiple root with an unknown multiplicity for the equation $f(k) = 0$ by transforming it into the problem of finding a simple root of an equivalent equation through the following transformation:

$$\Phi(k) = \frac{f(k)}{f'(k)}. \tag{15}$$

If α is a root of f of multiplicity m with $f(k) = (k-\alpha)^m h(k)$, then

$$\Phi(k) = \frac{(k-\alpha)^m h(k)}{m(k-\alpha)^{m-1}h(k) + (k-\alpha)^m h'(k)}, \tag{16}$$

$$= (k-\alpha)\frac{h(k)}{mh(k) + (k-\alpha)h'(k)}, \tag{17}$$

also has a root at α. However, $h(\alpha) \neq 0$, so

$$\frac{h(\alpha)}{mh(\alpha) + (\alpha-\alpha)h'(\alpha)} = \frac{1}{m} \neq 0, \tag{18}$$

and α is a simple root of $\Phi(k)$. Newton's method can then be applied to $\Phi(k)$ to give

$$g(k) = k - \frac{\Phi(k)}{\Phi'(k)} = k - \frac{f(k)/f'(k)}{\left\{[f'(k)]^2 - [f(k)]\left[f''(k)\right]\right\}/[f'(k)]^2}, \tag{19}$$

which simplifies to

$$g(k) = k - \frac{f(k)f'(k)}{[f'(k)]^2 - f(k)f''(k)}. \tag{20}$$

If g satisfies the necessary continuity conditions, using functional iteration on g will achieve quadratic convergence, regardless of the multiplicity of the root of f. Theoretically, the main drawback of this method is the additional computation needed for $f''(k)$ and the more complex process of calculating the iterates. Practically, however, multiple roots can lead to significant round-off errors because the denominator in Eq. (20) involves the difference between two numbers that are both close to 0.

Following Traub's approach [21], Parida and Gupta [22] introduced a scheme applicable to both cases of roots with known and unknown multiplicity. Petković et al. [23] proposed two classes of methods that handle both scenarios by using appropriate accelerating generators for iterative functions. Li et al. [24] developed a new fifth-order modified Newton's method designed to find multiple roots of nonlinear equations with unknown multiplicity. Additionally, Sharma and Bahl [25] presented a sixth-order modified Newton's method based on Traub's [21] transformation.

2.2 Optimal eighth-order iterative schemes for finding multiple roots with unknown multiplicity

Only two optimal eighth-order iterative methods have been developed for finding multiple roots of nonlinear equations with unknown multiplicity. In the year 2016, Jaiswal [26] claimed that asserted that he was the first to introduce an optimal eighth-order method for addressing multiple roots when the multiplicity is unknown. The method is presented as follows:

$$\begin{aligned} y_t &= k_t - \frac{f(k_t)}{g_1(k_t, z_t)}, \\ u_t &= y_t - \frac{f(y_t)}{g_2(k_t, z_t, y_t)}, \\ k_{t+1} &= u_t - \frac{f(u_t)}{g_3(k_t, z_t, y_t, u_t)}, \end{aligned} \tag{21}$$

where $g_1(k_t, z_t) = \frac{f(z_t) - f(k_t)}{f(k_t)}, z_t = k_t + f(k_t)$,

$$g_2(k_t, y_t, z_t) = \frac{f[k_t, y_t] f[y_t, z_t]}{f[k_t, z_t]}, g_3(k_t, z_t, y_t, u_t) = b_2 - b_1 b_4,$$

and

$$\begin{aligned} &b_1 = f(u_t), b_4 = \frac{f[y_t, u_t, k_t] - f[y_t, u_t, z_t]}{f[y_t, z_t] - f[y_t, k_t]}, b_3 = f[y_t, u_t, z_t] + b_4 f[y_t, z_t], \\ &b_2 = f[y_t, u_t] - b_3(y_t - u_t) + f(y_t) b_4. \end{aligned}$$

Theorem 1.2 Assume that $f \in \mathbb{C}^8(D)(D \subseteq \mathbb{R} \to \mathbb{R})$, and contains only one root $\alpha \in D$, where D is an open subset of $\mathbb{R}$ and s is sufficiently large. If the starting guess k_0 was near enough to α, the iterative scheme (21) achieves an optimal convergence rate of eight, and the error equation is given by:

$$\begin{aligned} e_{t+1} = &\frac{1}{m^{11}}\left[(1+m)^4 b_1\left((-1+m)b_1^2 - 2mb_2\right)\left(-(-1+m)b_1^4 + m(2+m)b_1^2 b_2 - 4m^2 b_2^2 + 3m^2 b_1 b_3\right)\right]e_t^8 \\ &+ O(e_t^9). \end{aligned} \tag{22}$$

In the year 2021, Sariman et al. [27] modified the Newton-Householder method by adopting Traub's [21] concept and approximating the weight function as a rational function. The optimal method proposed for finding multiple roots of unknown multiplicity was of eighth order with four functional evaluations at each iteration.

$$y_t = k_t - \frac{\Phi(k_t)}{\Phi'(k_t)}, t \geq 0,$$

$$z_t = y_t - \frac{\Phi(y_t)}{\Phi'(k_t)} - \frac{\Phi(y_t)^2}{2\Phi'^3(k_t)} \cdot \left(\frac{\alpha\Phi(y_t) + \beta\Phi(k_t)}{(y_t - k_t)^2} \right),$$

$$k_{t+1} = z_t - \frac{\Phi(z_t)}{\Phi'(k_t)} \cdot \left(\frac{p_0 + p_1 s + p_2 s^2 + p_3 s^3 + u(p_4 + p_5 s + p_6 s^2)}{q_0 + q_1 s + q_2 s^2 + q_3 s^3 + u(q_4 + q_5 s + q_6 s^2)} \right), \tag{23}$$

where $\Phi(k_t) = \frac{f(k_t)}{f'(k_t)}, s = \frac{\Phi(y_t)}{\Phi(k_t)}, u = \frac{\Phi(z_t)}{\Phi(y_t)}$ and $\alpha, \beta, p_t, q_t\ (0 \le t \le 6) \in \mathbb{R}$.

Theorem 1.3 Assume that $\alpha \in \mathbb{C}$ is the root of $f(k)$ with multiplicity m. If the initial value k_0 is sufficiently close to the root α, the iteration scheme (23) can reach a convergence order of eight when

$$\alpha = 10, \beta = 4, p_0 = q_0 = -8, p_2 = -16, p_3 = 25, q_1 = 16, q_3 = -23, q_4 = 8, p_1 = p_4 = p_5 = p_6 = q_2 = q_5 = q_6 = 0,$$

this yields the error equation:

$$e_{t+1} = \frac{1}{4m^7}\left[C_1^2\left(C_1^2(m+1) - 2C_2 m\right)\left(12C_3 m^2 + C_1^3(2m-3)(2m+7) - 4C_2 C_1 m(3m+4)\right)\right]e_t^8 + O\left(e_t^9\right). \tag{24}$$

Another set of parameter values that gives convergence of order eight is

$$\alpha = 10, \beta = 4, p_0 = q_0 = 1, p_1 = p_6 = 2, p_2 = p_3 = 6, p_4 = -1, q_4 = -2,\ p_5 = q_1 = q_2 = q_3 = q_5 = q_6 = 0,$$

which gives:

$$e_{t+1} = \frac{1}{m^5}\left[C_1\left(C_2\left(C_1^2 - 4C_2\right) + 3C_1C_3\right)\left(C_1^2(m+1) - 2C_2 m\right)\right]e_t^8 + O\left(e_t^9\right). \tag{25}$$

2.2.1 Special case of weight function (SA – 1)

We choose the special case of this method (23) and is given below:

$$y_t = k_t - \frac{\Phi(k_t)}{\Phi'(k_t)}, t \ge 0,$$

$$z_t = y_t - \frac{\Phi(y_t)}{\Phi'(k_t)} - \frac{\Phi(y_t)^2}{2\Phi'^3(k_t)} \cdot \left(\frac{10\Phi(y_t) + 4\Phi(k_t)}{(y_t - k_t)^2} \right),$$

$$k_{t+1} = z_t - \frac{\Phi(z_t)}{\Phi'(k_t)} \cdot \left(\frac{-8 - 16s^2 + 25s^3}{-8 + 16s - 23s^3 + 8u} \right), \tag{26}$$

where $\Phi(k_t) = \frac{f(k_t)}{f'(k_t)}, s = \frac{\Phi(y_t)}{\Phi(k_t)}, u = \frac{\Phi(z_t)}{\Phi(y_t)}$. We named it *SA* – 1.

Motivated by previous research, we have developed another eighth-order method for the case of unknown multiplicity. This new method is detailed below, accompanied by numerical results and dynamical analysis comparing it with existing methods for known and unknown multiplicity.

3. Construction of optimal eighth-order iterative scheme with unknown multiplicity

We have developed an efficient and simple eighth-order iterative family for finding multiple roots of nonlinear equations when the multiplicity is not known in advance. Our newly introduced method, outlined in Eq. (27), is based on King's method in the second step. In the third step, we utilize a uni-variate weight function along with divided differences of various orders, as detailed below:

$$y_t = k_t - \frac{\Phi(k_t)}{\Phi'(k_t)}, t \geq 0,$$

$$z_t = y_t - \frac{\Phi(k_t) + \beta\Phi(y_t)}{\Phi(k_t) + (\beta - 2)\Phi(y_t)} \cdot \frac{\Phi(y_t)}{\Phi'(k_t)},$$

$$k_{t+1} = z_t - H(u_t) \cdot \frac{\Phi(z_t)}{\Phi[z_t, y_t] + \Phi[z_t, k_t, k_t] \cdot (z_t - y_t)}, \tag{27}$$

where $\Phi(k) = \frac{f(k)}{f'(k)}, u_t = \frac{\Phi(z_t)}{\Phi(k_t)},.$

and divided difference is given as below:

$$\Phi[z_t, y_t] = \frac{\Phi(z_t) - \Phi(y_t)}{z_t - y_t}, \tag{28}$$

$$\Phi[z_t, k_t, k_t] = \frac{\Phi[z_t, k_t] - \Phi'(k_t)}{z_t - k_t}, \tag{29}$$

where $\Phi[z_t, k_t] = \frac{\Phi(z_t) - \Phi(k_t)}{z_t - k_t}$.

Theorem 1.4 Suppose that $\alpha \in \mathbb{C}$ is the root of $f(k)$ with multiplicity m if the initial estimate k_0 is sufficiently near to the root α the iterative approach (Eq. (27)) can attain the convergence order of eight when

$$H_0 = 1; H_1 = 2; \beta = -\frac{1}{2};$$

and the error equation is expressed as:

$$e_{t+1} = \frac{1}{m^7}\left\{C_1^2\left(-2C_2m + C_1^2(1+m)\right)\left(-3C_1C_2m^2 + 3C_3m^2 + C_1^3(-4+m^2)\right)e_t^8\right\} + O\left(e_t^9\right), \tag{30}$$

where $e_t = k_t - \alpha$ and $C_q = \frac{m!}{(m+q)!} \frac{f^{(m+q)}(\alpha)}{f^{(m)}(\alpha)}$ for $q \in \mathbb{N}$.

3.1 Some special cases of weight functions

In this part, we explore different weight schemes that satisfy the conditions stated in above theorem using our new method (27) and are provided as follows:

$$H(u_t) = H_0 + H_1 u_t + H_2 \frac{u_t^2}{2!} + H_3 \frac{u_t^3}{3!}. \tag{31}$$

We have following four cases described below:

3.1.1 Case 1

We consider the rational weight function $H(u_t)$, satisfying conditions $H_0 = 1; H_1 = 2;$ which can be expressed as follows:

$$H(u_t) = \frac{1 + 4u + 4u^2}{1 + 2u} = \frac{1 + 4u(1 + u)}{1 + 2u}. \tag{32}$$

SMFL – 1: We derive a unique case of our newly created scheme (27) of optimal eighth order of convergence for multiple roots in case of unknown multiplicity for this specific case of the weight function presented in (32) known as *SMFL* – 1, which is given by:

$$\begin{aligned} y_t &= k_t - \frac{\Phi(k_t)}{\Phi'(k_t)}, t \geq 0, \\ z_t &= y_t - \frac{2\Phi(k_t) - \Phi(y_t)}{2\Phi(k_t) - 5\Phi(y_t)} \cdot \frac{\Phi(y_t)}{\Phi'(k_t)}, \\ k_{t+1} &= z_t - \left(\frac{1 + 4u(1 + u)}{1 + 2u}\right) \cdot \frac{\Phi(z_t)}{\Phi[z_t, y_t] + \Phi[z_t, k_t, k_t] \cdot (z_t - y_t)}, \end{aligned} \tag{33}$$

where $\Phi(k) = \frac{f(k_t)}{f'(k_t)}, u_t = \frac{\Phi(z_t)}{\Phi(k_t)}$.

3.1.2 Case 2

We take the trignometric weight function $H(u_t)$, satisfying conditions $H_0 = 1; H_1 = 2;$ which can be represented as:

$$H(u_t) = 2u + \cos(u)^2. \tag{34}$$

SMFL – 2: To identify multiple roots of unknown multiplicity, we succeed in the unique scenario of our recently developed scheme (27) of optimal convergence order of eight, known as *SMFL* – 2, which is given by:

$$\begin{aligned} y_t &= k_t - \frac{\Phi(k_t)}{\Phi'(k_t)}, t \geq 0, \\ z_t &= y_t - \frac{2\Phi(k_t) - \Phi(y_t)}{2\Phi(k_t) - 5\Phi(y_t)} \cdot \frac{\Phi(y_t)}{\Phi'(k_t)}, \end{aligned}$$

$$k_{t+1} = z_t - \left(2u + \cos(u)^2\right) \cdot \frac{\Phi(z_t)}{\Phi\left[z_t, y_t\right] + \Phi[z_t, k_t, k_t] \cdot \left(z_t - y_t\right)}, \tag{35}$$

where $\Phi(k) = \frac{f(k_t)}{f'(k_t)}, u_t = \frac{\Phi(z_t)}{\Phi(k_t)}$.

3.1.3 Case 3

We consider another trignometric weight function $H(u_t)$, satisfying conditions $H_0 = 1; H_1 = 2;$ which can be expressed as follows:

$$H(u_t) = 2u + \cos(u)^2 + \sin(u)^2. \tag{36}$$

SMFL – 3: Using the unique weight function selection provided in Eq. (36), we obtain the newly created scheme (27) of optimal convergence order of eight for multiple roots in case of unknown multiplicity known as *SMFL* – 3, which is expressed as follows:

$$\begin{aligned} y_t &= k_t - \frac{\Phi(k_t)}{\Phi'(k_t)}, t \geq 0, \\ z_t &= y_t - \frac{2\Phi(k_t) - \Phi(y_t)}{2\Phi(k_t) - 5\Phi(y_t)} \cdot \frac{\Phi(y_t)}{\Phi'(k_t)}, \\ k_{t+1} &= z_t - \left(2u + \cos(u)^2 + \sin(u)^2\right) \cdot \frac{\Phi(z_t)}{\Phi\left[z_t, y_t\right] + \Phi[z_t, k_t, k_t] \cdot \left(z_t - y_t\right)}, \end{aligned} \tag{37}$$

where $\Phi(k) = \frac{f(k_t)}{f'(k_t)}, u_t = \frac{\Phi(z_t)}{\Phi(k_t)}$.

3.1.4 Case 4

We take the exponential weight function $H(u_t)$, satisfying conditions $H_0 = 1; H_1 = 2;$ can be represented as:

$$H(u_t) = u + \exp(u). \tag{38}$$

SMFL – 4: By using the weight function given in Eq. (38), the unique scenario of our recently developed optimal scheme (27) having eighth order for multiple roots with unknown multiplicity known as *SMFL* – 4 is given by:

$$\begin{aligned} y_t &= k_t - \frac{\Phi(k_t)}{\Phi'(k_t)}, t \geq 0, \\ z_t &= y_t - \frac{2\Phi(k_t) - \Phi(y_t)}{2\Phi(k_t) - 5\Phi(y_t)} \cdot \frac{\Phi(y_t)}{\Phi'(k_t)}, \\ k_{t+1} &= z_t - (u + \exp(u)) \cdot \frac{\Phi(z_t)}{\Phi\left[z_t, y_t\right] + \Phi[z_t, k_t, k_t] \cdot \left(z_t - y_t\right)}, \end{aligned} \tag{39}$$

where $\Phi(k) = \frac{f(k_t)}{f'(k_t)}, u_t = \frac{\Phi(z_t)}{\Phi(k_t)}$.

4. Numerical results

Our aim in the following portion is to examine the power, effectiveness and convergence behavior of the suggested iterative method (27) by using computational testing challenges and analyzing computing outcomes with previous methods having multiple roots. Comparison has been made on the basis of (t), $|f(k_t)|$, $|k_t - \alpha|$, $\eta = \left|\frac{k_{t+1}-k_t}{(k_t-k_{t-1})^8}\right|$ and the computational order of convergence COC. **Tables 1–6** compare the methods we proposed $SMFL-1$ to $SMFL-4$ with the methods that are already in usage $SF-2$ and $SA-1$ as stated in (14) and (26). Wolfram Mathematica 8 and Maple 18 have been employed for all computational tasks. We focused on multiple significant digits (minimum 1000) in our calculations to reduce round error; however, due to limitations, we could only show the results to four decimal places when displaying them in the form of tables. To determine the computational order of convergence, we used Jay's [28] formula, which is given below:

$$COC \approx \frac{\log|(k_{t+1}-\alpha)/(k_t-\alpha)|}{\log|(k_t-\alpha)/(k_{t-1}-\alpha)|}. \tag{40}$$

Furthermore, we have selected many of the problems and their precise root α, for which we compared the new technique with the existing methods. The real-world problems are listed below:

4.1 Beam designing problem

We examine the problem of beam positioning, where a beam of length p is directed precisely at the boundary of a cubical box, with each side measuring 1 unit in length. As illustrated in **Figure 1**, one end of the beam makes contact with the wall, while the other end rests on the floor.

Measuring the length of the floor between the bottom of the beam and the surface of the wall is a problem. Consider y is the length in measures over the bottom of the beam at the floor to the end of the box, and k is a length with units to the bottom of the beam at the floor toward the bottom of the box. After that using a particular measure for p, we obtain the nonlinear equation that follows:

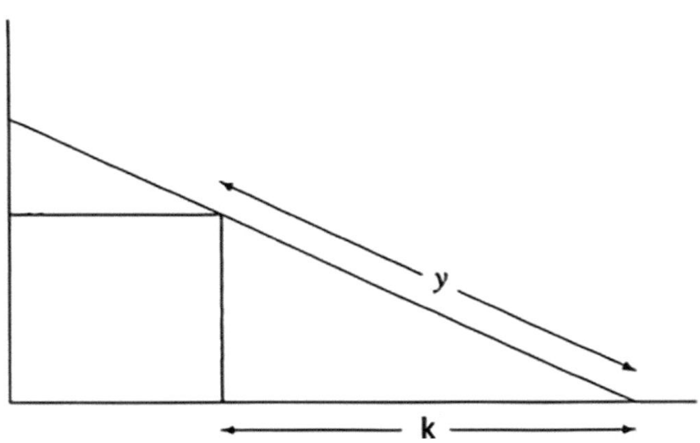

Figure 1.
Beam designing problem.

$f_1(k) = k^4 + 4k^3 - 24k^2 + 16k + 16, k = 2, m = 2, k_0 = 2.5$				
Methods	$k_1 - k_0$	$k_2 - k_1$	$k_3 - k_2$	COC
$SMFL-1$	9.7560×10^{-8}	2.3377×10^{-60}	2.5407×10^{-481}	8.00
$SMFL-2$	9.1517×10^{-8}	1.4016×10^{-60}	4.2429×10^{-483}	8.00
$SMFL-3$	9.7560×10^{-8}	2.3377×10^{-60}	2.5407×10^{-481}	8.00
$SMFL-4$	1.0058×10^{-7}	2.9843×10^{-60}	1.7921×10^{-480}	8.00
$SA-1$	3.2352×10^{-7}	9.9716×10^{-57}	8.1211×10^{-453}	7.99
$SF-2$	6.8121×10^{-6}	5.8044×10^{-44}	1.6128×10^{-348}	7.99

Table 1.
Results of $|k_n - k_{n-1}|$ of different methods for $f_1(k)$.

$f_1(k) = k^4 + 4k^3 - 24k^2 + 16k + 16, k = 2, m = 2, k_0 = 2.5$				
Methods	$f(k_1)$	$f(k_2)$	$f(k_3)$	η
$SMFL-1$	2.2843×10^{-13}	1.3116×10^{-118}	1.5492×10^{-960}	2.8483×10^{-4}
$SMFL-2$	2.0101×10^{-13}	4.7149×10^{-119}	4.3205×10^{-964}	2.8483×10^{-4}
$SMFL-3$	2.2843×10^{-13}	1.3116×10^{-118}	1.5492×10^{-960}	2.8483×10^{-4}
$SMFL-4$	2.4281×10^{-13}	2.1375×10^{-118}	7.7087×10^{-959}	2.8483×10^{-4}
$SA-1$	2.5120×10^{-12}	2.3864×10^{-111}	1.5828×10^{-903}	8.3075×10^{-5}
$SF-2$	1.1137×10^{-9}	8.0859×10^{-86}	6.2434×10^{-695}	1.2517×10^{-2}

Table 2.
Results of $|f(k_n)|$ of different methods for $f_1(k)$.

$$f_1(k) = k^4 + 4k^3 - 24k^2 + 16k + 16, \tag{41}$$

with a multiple root $k = 2.0$ of multiplicity $m = 2$. For this test function, we took $k_0 = 2.5$ and the comparison results are illustrated in **Tables 1** and **2**.

4.2 Root clustering challenge

We address the root clustering challenge, which is comparable to the problem studied by Zeng [29]. The specific function we consider is:

$$f_2(k) = (k-1)^{120}(k-2)^{150}(k-3)^{100}(k-4)^{55}. \tag{42}$$

The roots of f_2 are $k = 1,2,3$ and 4, with multiplicity $m = 120, 150, 100$ and 55, respectively. These multiplicities indicate that the roots are not simple but rather have high orders of repetition.

The roots are all relatively close together. Thus, this is referred to as a root cluster challenge. For the numerical study, we used $k = 4$ as the multiple root with multiplicity $m = 55$. **Tables 3** and **4** show the computational findings, using an initial approximation $k_0 = 4.1$.

$f_2(k) = (k-1)^{120}(k-2)^{150}(k-3)^{100}(k-4)^{55}, k = 4, m = 55, k_0 = 4.1$				
Methods	$k_1 - k_0$	$k_2 - k_1$	$k_3 - k_2$	COC
$SMFL-1$	2.1578×10^{-4}	3.0745×10^{-25}	5.2051×10^{-192}	8.00
$SMFL-2$	1.9623×10^{-4}	1.4378×10^{-25}	1.1907×10^{-194}	8.00
$SMFL-3$	2.1578×10^{-4}	3.0745×10^{-25}	5.2051×10^{-192}	8.00
$SMFL-4$	2.2579×10^{-4}	4.4192×10^{-25}	9.4827×10^{-191}	8.00
$SA-1$	1.6644×10^{-3}	4.4524×10^{-18}	1.2332×10^{-134}	7.99
$SF-2$	4.7122×10^{-4}	4.7385×10^{-21}	5.0615×10^{-157}	7.99

Table 3.
Results of $|k_n - k_{n-1}|$ *of different methods for* $f_2(k)$.

$f_2(k) = (k-1)^{120}(k-2)^{150}(k-3)^{100}(k-4)^{55}, k = 4, m = 55, k_0 = 4.1$				
Methods	$f(k_1)$	$f(k_2)$	$f(k_3)$	η
$SMFL-1$	5.7612×10^{-100}	1.7257×10^{-1246}	$6.4977 \times 10^{-10419}$	6.5395×10^{4}
$SMFL-2$	3.1182×10^{-102}	1.2099×10^{-1264}	$3.7940 \times 10^{-10564}$	6.5376×10^{4}
$SMFL-3$	5.7612×10^{-100}	1.7257×10^{-1246}	$6.4977 \times 10^{-10419}$	6.5395×10^{4}
$SMFL-4$	6.9565×10^{-99}	7.9975×10^{-1238}	$1.3819 \times 10^{-10349}$	6.5404×10^{4}
$SA-1$	2.6477×10^{-51}	1.2077×10^{-852}	2.6158×10^{-7263}	7.5598×10^{4}
$SF-2$	3.0254×10^{-81}	3.7108×10^{-1016}	1.3950×10^{-8494}	1.9489×10^{6}

Table 4.
Results of $|f(k_n)|$ *of different methods for* $f_2(k)$.

4.3 Osteoporosis in Chinese women

The following equation studies the variation in age-related speed of sound (S) at the tibia and the prevalence of osteoporosis in Chinese women [30], where k is the age in years:

$$S(k) = 0.004k^3 - 0.78k^2 + 39.9k - 3383. \tag{43}$$

Here, S is measured in m/s. For a research subject with S measured as 3995.5 m/s, we can determine the most likely age of the woman using the equation:

$$f_3(k) = 0.004k^3 - 0.78k^2 + 39.9k - 612.5. \tag{44}$$

The roots of f_3 are $k = 35, 35$ and 125. For the numerical study, we used $k = 35$ as the multiple root of multiplicity $m = 2$ using an initial approximation $k_0 = 34.5$.

For the test problem $f_1(k)$, **Tables 1** and **2** demonstrate that the convergence behavior of the newly presented methods is superior to that of existing methods. Specifically, our methods exhibit reduced absolute error and asymptotic error constants compared to the already established techniques $SA-1$ and $SF-2$. Among these, the $SMFL-4$ approach outperforms all other methods, including $SF-2$ and $SA-1$.

$f_3(k) = 0.004k^3 - 0.78k^2 + 39.9k - 612.5, k = 35, m = 2, k_0 = 34.5$				
Methods	$k_1 - k_0$	$k_2 - k_1$	$k_3 - k_2$	COC
$SMFL - 1$	1.0003×10^{-19}	5.4602×10^{-188}	1.0971×10^{-377}	8.99
$SMFL - 2$	1.0015×10^{-19}	5.5212×10^{-188}	7.4295×10^{-378}	8.99
$SMFL - 3$	1.0003×10^{-19}	5.4602×10^{-188}	1.0971×10^{-377}	8.99
$SMFL - 4$	9.9969×10^{-20}	5.4299×10^{-188}	1.2202×10^{-377}	8.99
$SA - 1$	5.2361×10^{-18}	7.6141×10^{-154}	4.8197×10^{-378}	7.99
$SF - 2$	1.4961×10^{-16}	1.0209×10^{-140}	1.0540×10^{-498}	D

Table 5.
Results of $|k_n - k_{n-1}|$ of different methods for $f_1(k)$.

$f_3(k) = 0.004k^3 - 0.78k^2 + 39.9k - 612.5, k = 35, m = 2, k_0 = 34.5$				
Methods	$f(k_1)$	$f(k_2)$	$f(k_3)$	η
$SMFL - 1$	3.6022×10^{-39}	1.0733×10^{-375}	4.3337×10^{-755}	5.4463×10^{-36}
$SMFL - 2$	3.6111×10^{-39}	1.0974×10^{-375}	1.9871×10^{-755}	5.4531×10^{-36}
$SMFL - 3$	3.6022×10^{-39}	1.0733×10^{-375}	4.3337×10^{-755}	5.4463×10^{-36}
$SMFL - 4$	3.5978×10^{-39}	1.0614×10^{-375}	5.3606×10^{-756}	5.4430×10^{-36}
$SA - 1$	9.8701×10^{-36}	2.0871×10^{-307}	8.3627×10^{-756}	1.3475×10^{-15}
$SF - 2$	8.0589×10^{-33}	3.7521×10^{-281}	0	4.0651×10^{-14}

Table 6.
Results of $|f(k_n)|$ of different methods for $f_1(k)$.

For the test function $f_2(k)$, **Tables 3** and **4** reveal that the recently introduced approaches $SMFL - 1$ to $SMFL - 4$ surpass the performance of the existing methods $SA - 1$ and $SF - 2$. For the test function $f_3(k)$, $SMFL - 2$ shows the lowest absolute error and $SF - 2$ diverges for computational order of convergence. The newly proposed methods consistently outperform the established methods $SA - 1$ and $SF - 2$, as evidenced **Tables 1–6**.

5. Dynamical analysis

Our objective is to assess and compare the effectiveness of the proposed methods by utilizing a graphical tool called the basins of attraction. This concept, first introduced by Vrscay and Gilbert [31], helps in analyzing the convergence and dynamical behavior of iterative schemes. To assess stability, we plot the dynamical planes for each method $SMFL - 1$, $SMFL - 2$, $SA - 1$, and $SF - 2$, when applied to the nonlinear functions $f_1(k) - f_3(k)$. The complex dynamical planes, depicted in **Figures 2–7**, were generated following the procedure outlined in [32].

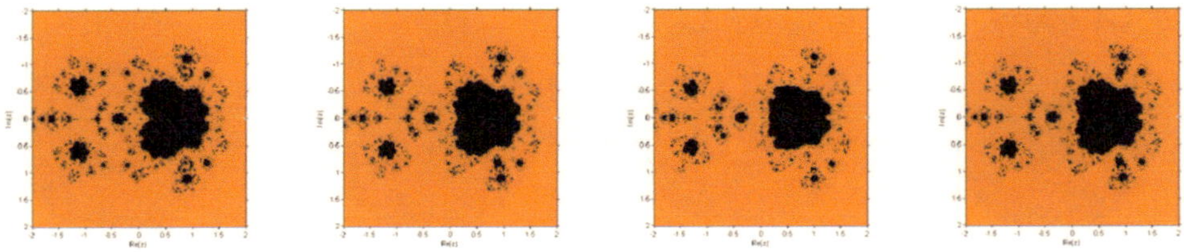

(a) $SMFL-1$ (b) $SMFL-2$ (c) $SMFL-3$ (d) $SMFL-4$

Figure 2.
Graphics of $SMFL - 1$ *to* $SMFL - 4$ *for* $f_1(k)$.

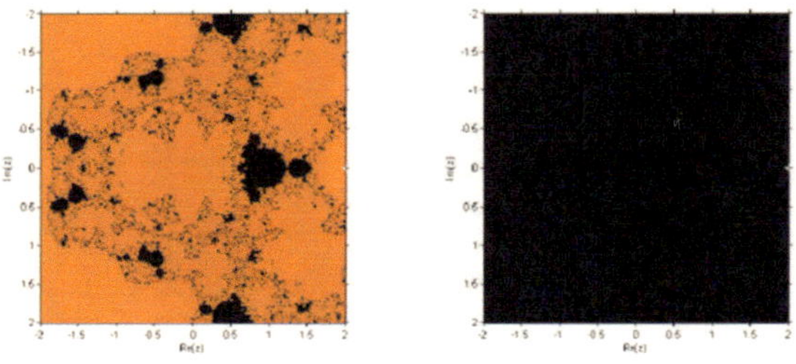

(a) $SA-1$ (b) $SF-2$

Figure 3.
Graphics of $SA - 1$ *and* $SF - 2$ *for* $f_1(k)$.

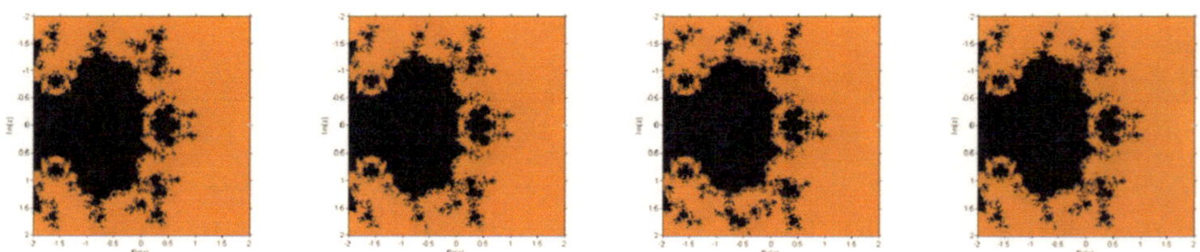

(a) $SMFL-1$ (b) $SMFL-2$ (c) $SMFL-3$ (d) $SMFL-4$

Figure 4.
Graphics of $SMFL - 1$ *to* $SMFL - 4$ *for* $f_2(k)$.

To compare the stability of the methods, a grid of $[500 \times 500]$ points is generated in the complex plane to evaluate how extensively the initial guesses lead to the intended repeated root. The process involves up to 80 iterations, with a stopping criterion set at 10^{-3}. If the sequence of iterations converges to the multiple zero, the corresponding region is highlighted in orange. Conversely, the region where the sequence fails to converge to the multiple root or converges to unintended fixed points (which are not roots of the nonlinear function) is shown in black. A white star marks the location of the multiple zero in **Figures 2–7**.

Figures 2–7 depict the convergence and divergence regions for the new schemes $SMFL-1$ to $SMFL-4$, in comparison with other schemes of the same order. The

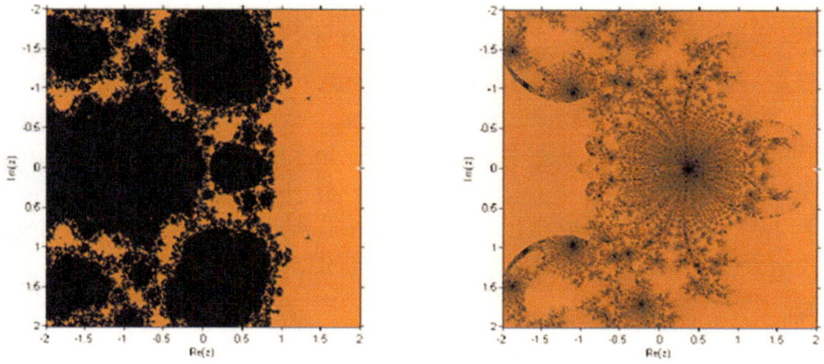

(a) $SA-1$ (b) $SF-2$

Figure 5.
Graphics of SA − 1 and SF − 2 for $f_2(k)$.

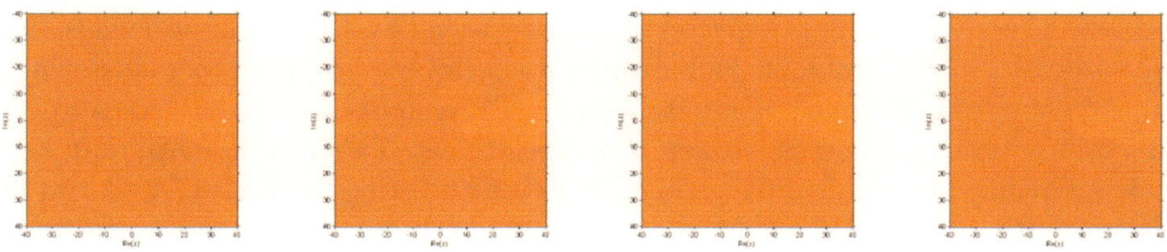

(a) $SMFL-1$ (b) $SMFL-2$ (c) $SMFL-3$ (d) $SMFL-4$

Figure 6.
Graphics of SMFL − 1 to SMFL − 4 for $f_3(k)$.

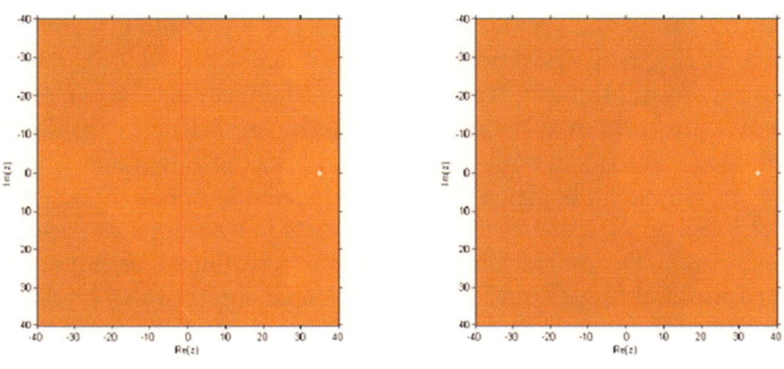

(a) $SA-1$ (b) $SF-2$

Figure 7.
Graphics of SA − 1 and SF − 2 for $f_3(k)$.

results show that the $SMFL-1$ to $SMFL-4$ methods demonstrate a broader convergence region compared to the earlier schemes $SA-1$ and $SF-2$. For $f_1(k)$, the new schemes $SMFL-1$ to $SMFL-4$ demonstrate greater stability compared to $SA-1$ and $SF-2$, as $SF-2$ shows a predominantly divergent region. For $f_2(k)$, the divergence region of $SA-1$ is more extensive. For the function $f_3(k)$, all methods exhibit minimal divergence and converge more rapidly within their shared regions of convergence. This highlights the stable and consistent performance of the newly developed schemes, as evident from the graphical representations.

6. Conclusions

In this chapter, we have developed a unique three-step family of iterative techniques to solve nonlinear equations with multiple roots of unknown multiplicity. Utilizing the weight function technique and divided differences of various orders, we have devised an optimal eighth-order convergent scheme. A thorough convergence analysis, which guarantees eighth-order convergence, was conducted. This analysis requires three function evaluations and one derivative evaluation per complete cycle, resulting in an efficiency index of $8^{\frac{1}{4}} = 1.6818$.

A few effective and simple special cases have been developed and utilized to deal with some real-world problems including beam designing problem, root clustering challenge, and osteoporosis in Chinese women. These applications allowed us to compare numerical outcomes with existing approaches. **Tables 1–6** reveal that our newly introduced methods consistently outperform existing methods ($SA-1$ and $SF-2$) by exhibiting reduced absolute error and lower asymptotic error constants. Among these, the $SMFL-4$ approach outperforms all other methods.

Basins of attraction are utilized as dynamic techniques to visualize the stability of competitive iterative methods. **Figures 2–7** illustrate that $SMFL-1$ to $SMFL-4$ methods have a wider convergence region and greater stability compared to earlier schemes. These graphics clearly demonstrate the stability and consistency of the newly developed schemes.

The methods discussed in this chapter indicate that our newly created schemes are more reliable, effective, and robust within their respective fields. The results, supported by the tables and graphics, affirm the superiority and efficiency of our approaches in solving real-world problems.

The future scope of "Iterative Techniques for Nonlinear Equations: Addressing Multiple Roots with Unknown Multiplicity" is highly promising, with significant potential for advancements in both theoretical and practical domains. These advanced high-order iterative methods play a critical role in solving nonlinear equations with multiple roots, delivering substantial improvements in convergence speed and computational efficiency. Future research may focus on refining algorithms, conducting complex dynamic analyses, expanding computational applications, implementing robust software solutions, and promoting interdisciplinary collaboration.

Conflict of interest

The authors declare no conflict of interest.

Author details

Saima Akram[1,2]†, Moin-Ud-Din Junjua[3,4*]†, Faiza Akram[2]† and Laiba Khalid[1]†

1 Department of Mathematics, Government College Women University Faisalabad, Faisalabad, Pakistan

2 Centre for Advanced Studies in Pure and Applied Mathematics, Bahauddin Zakariya University, Multan, Pakistan

3 School of Mathematical Sciences, Zhejiang Normal University, Jinhua, China

4 Department of Mathematics, Ghazi University, Dera Ghazi Khan, Pakistan

*Address all correspondence to: moinuddin@zjnu.edu.cn

† These authors contributed equally.

References

[1] Richard L, Burden J. Douglas Faires. Numerical Analysis. 2011;**7**:47-101

[2] Bradie B. A Friendly Introduction to Numerical Analysis. India: Pearson Education; 2006

[3] Ostrowski AM. Solution of Equations and Systems of Equations: Pure and Applied Mathematics: A Series of Monographs and Textbooks. Vol. 9. London: Academic Press Inc. (London) Ltd; 2016

[4] Kung HT, Traub JF. Optimal order of one-point and multipoint iteration. Journal of the ACM (JACM). 1974;**21**(4): 643-651

[5] Rall LB. Convergence of the Newton process to multiple solutions. Numerische Mathematik. 1966;**9**:23-37

[6] Schröder E. Über unendlich viele Algorithmen zur Auflösung der Gleichungen. Mathematische Annalen. 1870;**2**(2):317-365

[7] Shams M, Carpentieri B. Q-analogues of parallel numerical scheme based on neural networks and their engineering applications. Applied Sciences. 2024; **14**(4):1540

[8] Shams M, Carpentieri B. Efficient inverse fractional neural network-based simultaneous schemes for nonlinear engineering applications. Fractal and Fractional. 2023;**7**(12):849

[9] Shams M, Kausar N, Agarwal P, Edalatpanah SA. Fractional Caputo-type simultaneous scheme for finding all polynomial roots. In: Recent Trends in Fractional Calculus and its Applications. Academic Press, an Imprint of Elsevier Science; 2024. pp. 261-272

[10] Rafiq N, Akram S, Shams M, Mir NA. Computer geometries for finding all real zeros of polynomial equations simultaneously. Computer Material and Continua. 2021;**69**(2): 2636-2651

[11] Shams M, Kausar N, Araci S, Oros GI. Artificial hybrid neural network-based simultaneous scheme for solving nonlinear equations: Applications in engineering. Alexandria Engineering Journal. 2024;**1**(108): 292-305

[12] Mir NA, Shams M, Rafiq N, Akram S, Rizwan M. Derivative free iterative simultaneous method for finding distinct roots of polynomial equation. Alexandria Engineering Journal. 2020;**59**(3):1629-1636

[13] Chun C, Neta B. A third-order modification of Newton's method for multiple roots. Applied Mathematics and Computation. 2009;**211**(2):474-479

[14] Mir NA, Anwar M, Shams M, Rafiq N, Akram S. On numerical schemes for determination of all roots simultaneously of non-linear equation. Mehran University Research Journal of Engineering and Technology. 2022; **41**(2):208-218

[15] Chun C, Ju Bae H, Neta B. New families of nonlinear third-order solvers for finding multiple roots. Computers & Mathematics with Applications. 2009; **57**(9):1574-1582

[16] Akram S, Zafar F, Yasmin N. An optimal eighth-order family of iterative methods for multiple roots. Mathematics. 2019;**7**(8):672

[17] Neta B. Extension of Murakami's high-order non-linear solver to multiple roots. International Journal of Computer Mathematics. 2010;**87**(5):1023-1031

[18] Behl R, Cordero A, Motsa SS, Torregrosa JR, Kanwar V. An optimal fourth-order family of methods for multiple roots and its dynamics. Numerical Algorithms. 2016;**71**:775-796

[19] Singh A, Jaiswal JP. An efficient family of optimal fourth-order iterative methods for finding multiple roots of nonlinear equations. Proceedings of the National Academy of Sciences, India Section A: Physical Sciences. 2015;**85**:439-450

[20] Akram S, Akram F, Junjua MU, Arshad M, Afzal T. A family of optimal eighth order iteration functions for multiple roots and its dynamics. Journal of Mathematics. 2021;**2021**(1):5597186

[21] Traub JF. Iterative Methods for the Solution of Equations. New York, N.Y.: Chelsea Publishing Company; 1982

[22] Parida PK, Gupta DK. An improved method for finding multiple roots and it's multiplicity of nonlinear equations in R. Applied Mathematics and Computation. 2008;**202**(2):498-503

[23] Petković MS, Petković LD, Džunić J. Accelerating generators of iterative methods for finding multiple roots of nonlinear equations. Computers & Mathematics with Applications. 2010; **59**(8):2784-2793

[24] Li X, Mu C, Ma J, Hou L. Fifth-order iterative method for finding multiple roots of nonlinear equations. Numerical Algorithms. 2011;**57**:389-398

[25] Sharma R, Bahl A. A sixth order transformation method for finding multiple roots of nonlinear equations and basin attractors for various methods. Applied Mathematics and Computation. 2015;**15**(269):105-117

[26] Jaiswal JP. An optimal order method for multiple roots in case of unknown multiplicity. Algorithms. 2016;**9**(1):10

[27] Sariman SA, Hashim I, Samat F, Alshbool M. Modification of Newton-Househölder method for determining multiple roots of unknown multiplicity of nonlinear equations. Mathematics. 2021;**9**(9):1020

[28] Jay LO. A note on Q-order of convergence. BIT Numerical Mathematics. 2001;**41**:422-429

[29] Zeng Z. Computing multiple roots of inexact polynomials. Mathematics of Computation. 2005;**74**(250):869-903

[30] King MR, Mody NA. Numerical and Statistical Methods for Bioengineering: Applications in MATLAB. New York, United States of America, Cambridge, UK: Cambridge University Press; 2010

[31] Vrscay ER, Gilbert WJ. Extraneous fixed points, basin boundaries and chaotic dynamics for Schröder and König rational iteration functions. Numerische Mathematik. 1987;**52**:1-6

[32] Chicharro FI, Cordero A, Torregrosa JR. Drawing dynamical and parameters planes of iterative families and methods. The Scientific World Journal. 2013;**2013**(1):780153

Chapter 4

Stability Analysis of Fractional Order Schemes for Nonlinear Engineering Problems

Mudassir Shams and Bruno Carpentieri

Abstract

Traditional analytical methods often struggle to handle the inherent complexity of nonlinear fractional problems, highlighting the need for robust numerical techniques. This chapter presents the formulation and stability assessment of novel high-order single-step iterative algorithms developed to solve nonlinear fractional equations arising in scientific and engineering applications. A detailed convergence analysis demonstrates that the proposed schemes achieve a convergence order of $(\delta + 1)$. Through parametric and dynamical analysis, we identify optimal parameter values that ensure stability, consistency, and computational efficiency, with particular attention to sensitivity to the initial guess. The performance of the proposed methods is evaluated against existing techniques in terms of residual error, error plots, computational order, and CPU time on selected real-world engineering problems.

Keywords: fractional order scheme, convergence analysis, stability analysis, dynamical planes, engineering applications

1. Introduction

About 1695, Leibniz and L'Hôpital introduced the concept of the semi-derivative, an early form of differentiation that laid the groundwork for the development of both classical and fractional calculus. Fractional calculus [1–3] extends the classical principles of differentiation and integration to non-integer orders, thereby enabling more accurate representations and a deeper understanding of complex phenomena characterized by memory effects, long-range dependence, and anomalous behaviors. This extension is particularly significant in scientific domains where systems exhibit nonlocal features, such as physics [4], engineering [5], biology [6], and finance. In fields including viscoelastic materials [7], electrochemistry [8], signal processing [9], and control theory [10], traditional integer-order models often fail to capture essential system dynamics. Fractional-order differential equations, derived from fractional calculus, offer a powerful framework for more accurately modeling and predicting the behavior of such systems [11–14].

Newton's method [15, 16], developed in the late seventeenth century by Isaac Newton and Joseph Raphson, has been widely used for many years to solve nonlinear equations

$$f(\epsilon) = 0. \tag{1}$$

It iteratively improves an initial estimate using a sequence of linear approximations of the form

$$\epsilon_{i+1} = \epsilon_i - \frac{f(\epsilon_i)}{f'(\epsilon_i)}, \tag{2}$$

that converge quickly—often at a quadratic order—to the actual solution. Various researchers have produced modified versions of Newton's method using different methodologies to construct efficient schemes for solving Eq. (1). For example, Tibor et al. [17] modified the Newton-type method using the geometric mean; Noor et al. [18] employed the Homotopy and Homotopy Perturbation Method to develop new variants; Özban introduced new versions of Newton's method [19]; Mir et al. [20] developed quadrature-based iterative methods; Chun [21] utilized the Decomposition Method; and Abbasbandy [22] applied the Modified Adomian Decomposition Method to enhance Newton-type methods. Additionally, the weight function technique has proven to be an efficient method for developing optimal one-step [23–26] and multi-step schemes [27, 28] for finding simple roots of Eq. (1).

Classic iterative approaches, such as the Newton-Raphson method, are widely used but exhibit limitations related to convergence rate, sensitivity to initial starting values, and high memory requirements. To overcome these limitations, researchers have increasingly incorporated fractional calculus into numerical algorithms. In recent years, several fractional-order iterative strategies have been developed to enhance the stability and convergence of root-finding methods by leveraging the memory and non-local properties of fractional derivatives. For example, Farman et al. [29] proposed a fractional-order Newton-Raphson approach that employs Caputo derivatives to evaluate stability and convergence. Similar efforts to evaluate the effectiveness of fractional-order iterative methods in nonlinear models have been reported by Naveen et al. [30], Ali et al. [31], and Erfanifar et al. [32], where fractional-order extensions of classical approaches have demonstrated enhanced numerical robustness. For additional examples and related approaches, see, examples [33–36] and references therein.

The development of higher-order formulations and the extension of Newton-type methods have been the focus of several further contributions. For example, Chang et al. [37] proposed a modified fractional Newton solver, while Shams and Carpentieri [38] developed a parallel fractional iterative scheme. In addition, Chicharro et al. [39] investigated the convergence and stability of second-order root-finders, laying the groundwork for visual and dynamical analyses of iterative algorithms.

Motivated by these considerations, we propose a novel, efficient, and stable single-step fractional-order scheme based on a generalized fractional Taylor expansion. We also employ modern visual tools to analyze its local and global convergence properties and to determine the optimal parameter values that further accelerate convergence. The main contributions of this chapter are as follows:

- The generalized fractional Taylor expansion and the Caputo derivative are employed to develop a new fractional-order iterative technique for solving nonlinear models.

- A rigorous local convergence analysis demonstrates that the scheme achieves an order of convergence equal to $\delta + 1$, where δ denotes the fractional parameter of the Caputo derivative.

- A detailed stability framework is presented to describe the scheme's convergence behavior at fixed points, strange fixed points, and infinity.

- Visual diagnostics based on parametric and dynamical planes are utilized to examine the global behavior of the method and to identify optimal parameter values that ensure robust convergence.

- Fractal-based graphics and numerical experiments are used to demonstrate the efficiency and superiority of the proposed iterative technique compared to existing methods.

To enhance convergence, the proposed method integrates fractional calculus in a novel way by developing iterative schemes based on the generalized fractional Taylor expansion. To the best of our knowledge, it is the first approach to combine fractional-order dynamics with visual stability tools—specifically, parametric and dynamical planes—for method optimization and analysis.

The remaining structure of the chapter is as follows. Section 2 constructs and analyzes the local convergence order of the fractional scheme. In Section 3, a dynamical analysis evaluates the stability of fractional-order iterative methods, including parametric and dynamical plane assessments. In Section 4, we apply the findings from Section 3 to solve two engineering problems using fractal analysis, demonstrating the effectiveness of the newly developed methods. The final section summarizes the research and presents concluding remarks.

2. Theoretical preliminaries

To provide clear context for the analysis presented in this paper, we first introduce the key terminology, notations, and concepts used throughout.

2.1 Gamma function

The Gamma function, also known as the generalized factorial function [40], is defined as follows:

$$\Gamma(\epsilon) = \int_0^{+\infty} u^{\varepsilon-1} e^{-u} du, \tag{3}$$

where $\epsilon > 0$. It satisfies $\Gamma(1) = 1$ and $\Gamma(n+1) = n!$ for $n \in N$.

2.2 Caputo fractional derivative

For a function

$$f : \mathbb{R} \to \mathbb{R}, \text{with} f \in \mathbb{C}^{+\infty}([\delta, \epsilon]), \text{where} -\infty < \delta < \epsilon < +\infty, \delta \geq 0, \text{and } m = [\delta] + 1, \tag{4}$$

the Caputo fractional derivative [41] of order δ is defined as:

$$\left[{}_{\breve{c}}\eth^{\delta}_{\delta_1}\right]f(\epsilon) = \begin{cases} \dfrac{1}{\Gamma(m-\delta)}\displaystyle\int_0^{\epsilon}\dfrac{d^m}{dt^m}f(t)\dfrac{1}{(\epsilon-t)^{\delta-m+1}}dt, & \delta\in N, \\ \dfrac{d^{m-1}}{dt^{m-1}}f(\epsilon), & \delta = m-1\in N\cup\{0\}, \end{cases} \tag{5}$$

where $\Gamma(\epsilon)$ is the Gamma function with $\epsilon > 0$.

Among various types of fractional derivatives, the Caputo derivative is particularly noteworthy because it satisfies the initial condition $\left[{}_{\breve{c}}\eth^{\delta}_{\delta_1}\right](C) = 0$ if δ is not a natural number and $C \in \mathbb{R}$. Given this property, we will focus on the Caputo derivative and introduce the fractional iterative approach for solving nonlinear equations using Caputo-type derivatives.

2.3 Generalized fractional Taylor expansion

Suppose $\left[{}_{\breve{c}}\eth^{\gamma\delta}_{\delta_1}\right]f(\epsilon)\in\breve{c}([\delta_1,\delta_2])$ for $\gamma = 1, \ldots, n+1$ where $\delta\in(0,1]$. Then the Generalized Taylor Formula [41] is given by

$$f(\epsilon) = \sum_{i=0}^{n}\left[{}_{\breve{c}}\eth^{i\delta}_{\delta 1}\right]f(\delta_1)\frac{(\epsilon-\delta_1)^{i\delta}}{\Gamma(i\delta+1)} + \left[{}_{\breve{c}}\eth^{(n+1)\delta}_{\delta_1}\right]f(\xi)\frac{(\epsilon-\delta_1)^{(n+1)\delta}}{\Gamma((n+1)\delta+1)}, \tag{6}$$

where

$$\delta_1 \le \xi \le \epsilon, \forall\epsilon\in(\delta_1,\delta_2]. \tag{7}$$

Additionally,

$$\left[{}_{\breve{c}}\eth^{n\delta}_{\delta_1}\right] = \left[{}_{\breve{c}}\eth^{\delta}_{\delta_1}\right]\cdot\left[{}_{\breve{c}}\eth^{\delta}_{\delta_1}\right]\ldots\left[{}_{\breve{c}}\eth^{\delta}_{\delta_1}\right](n\text{ times}). \tag{8}$$

Consider the Caputo-type Taylor expansion of $f(\epsilon)$ near $\delta_1 = \xi$ as:

$$f(\epsilon) = \frac{\left[{}_{\breve{c}}\eth^{1\delta}_{\xi}\right]f(\xi)}{\Gamma(+1)}(\epsilon-\xi)^{\delta} + \frac{\left[{}_{\breve{c}}\eth^{2\delta}_{\xi}\right]f(\xi)}{\Gamma(2\delta+1)}(\epsilon-\xi)^{2\delta} + O(\epsilon-\xi)^{3\delta}. \tag{9}$$

Taking $\frac{\left[{}_{\breve{c}}\eth^{1\delta}_{\xi}\right]f(\xi)}{\Gamma(\delta+1)}$ as common, we have:

$$f(\epsilon) = \frac{\left[{}_{\breve{c}}\eth^{1\delta}_{\xi}\right]f(\xi)}{\Gamma(\delta+1)}\left[(\epsilon-\xi)^{\delta} + \breve{c}_2(\epsilon-\xi)^{2\delta}\right] + O(\epsilon-\xi)^{3\delta}, \tag{10}$$

where

$$\breve{c}_j = \frac{\Gamma(\delta+1)}{\Gamma(\gamma\delta+1)}\frac{\left[{}_{\breve{c}}\eth^{\gamma\delta}_{\xi}\right]f(\xi)}{\left[{}_{\breve{c}}\eth^{\delta}_{\xi}\right]f(\xi)}, \gamma = 2, 3, \ldots \tag{11}$$

The corresponding Caputo-type derivative of $f(\varepsilon)$ around ξ is

$$\left[{}_{\breve{c}}\mathfrak{D}_{\xi}^{\delta}\right]f(\epsilon) = \frac{\left[{}_{\breve{c}}\mathfrak{D}_{\xi}^{1\delta}\right]f(\xi)}{\Gamma(\delta+1)}\left[\Gamma(\delta+1) + \frac{\Gamma(2\delta+1)}{\Gamma(\delta+1)}\breve{c}_2(\epsilon-\xi)^{\delta}\right] + O(\epsilon-\xi)^{2\delta}. \tag{12}$$

These results are used in the convergence analysis of our method.

2.4 Consistency

A root-finding technique is considered consistent if it preserves the fixed-point structure of the problem. Specifically, the condition $f(\xi) = \xi$ must hold for the iterative map $\epsilon_{i+1} = f(\epsilon_i)$, where ξ is a root of the nonlinear equation $f(\epsilon) = 0$. This property ensures that the technique accurately reflects the solution structure.

2.5 Stability concepts in fractional iterative methods

Stability is central to understanding the convergence behavior of an iterative technique. In this context, we distinguish three important types:

- *Numerical stability:* This refers to the method's sensitivity to small perturbations in the initial assumptions or round-off errors, as well as the propagation of errors during the iterative process.
- *Stability at fixed points:* This describes the convergence of the iterative map at fixed points. If $|f'(\xi)| < 1$, the fixed point is attracting (stable); if $|f'(\xi)| > 1$, it is repelling (unstable); and if $|f'(\xi)| = 1$, it is neutral.
- *Stability at infinity:* This refers to the method's global behavior for large-magnitude initial guesses. If the iterations diverge or exhibit chaotic trajectories, the approach is considered unstable near infinity.

Visual aids such as parametric and dynamical planes, which provide geometric insight into convergence patterns and global stability characteristics, are used to further illustrate these concepts in the following sections.

3. Fractional scheme construction and analysis

The construction of iterative methods for finding the roots of nonlinear equations can be significantly improved by using weight functions. Weight functions offer several benefits that enhance the efficiency and robustness of these methods in both conventional and fractional calculus. By carefully designing weight functions, the convergence rate of iterative methods can be greatly accelerated, often achieving super-linear or quadratic convergence [42]. In addition, weight functions improve the method's convergence rate by reducing numerical errors and dampening oscillations that may otherwise cause divergence. Moreover, they can help minimize computational effort by lowering the number of function evaluations needed for each iteration and enabling dynamic adaptation based on the progress of iterations. Due to these combined benefits, weight functions can be a powerful tool to use in the numerical analysis of nonlinear equations.

There has been little research on constructing fractional iterative schemes utilizing the fractional-weight function technique, which prompted this study. We propose the following variant of (2), and we refer to this method as BB^{δ} :

$$\epsilon_{i+1} = \epsilon_i - \left(\Gamma(\delta+1) \S \left(\frac{f(\epsilon_i)}{\left[{}_{c}^{\smile}\partial_{\delta_1}^{\delta} \right] f(\epsilon_i)} \right) \right)^{1/\delta}. \tag{13}$$

3.1 Convergence analysis

For the iterative scheme (13), we prove the following theorem to establish its order of convergence.

Theorem 1. *Let*

$$f : \partial \subseteq \mathbb{R} \to \mathbb{R} \tag{14}$$

be a continuous function with $\left[{}_{c}^{\smile}\partial_{\delta_1}^{\gamma\delta} \right] f(\epsilon)$ of order $\gamma\delta$ for any $\gamma \geq 0$ and $\delta \in (0, 1]$, containing the exact root ξ of $f(\epsilon)$. Furthermore, for a sufficiently close initial value ϵ_0, the convergence order of the Caputo-type fractional iterative schemes given by

$$\epsilon_{i+1} = \epsilon_i - \left(\Gamma(\delta+1) \S \left(\frac{f(\epsilon_i)}{\left[{}_{c}^{\smile}\partial_{\delta_1}^{\delta} \right] f(\epsilon_i)} \right) \right)^{1/\delta}, \tag{15}$$

is at least $\delta + 1$, and the error equation is as:

$$e_i^{[*]} = \left(\frac{(2^{\delta})^2 \Gamma(\delta + \frac{1}{2}) \breve{c}_2^2}{\delta \Gamma(\delta)\sqrt{\pi}} - \breve{c}_2^2 \right) e_i^{2\delta+1} + O\left(e_i^{3\delta+1}\right), \tag{16}$$

where

$$\breve{c}_j = \frac{\Gamma(\delta+1)}{\Gamma(\gamma\delta+1)} \frac{\left[{}_{c}^{\smile}\partial_{\xi}^{\gamma\delta} \right] f(\xi)}{\left[{}_{c}^{\smile}\partial_{\xi}^{\delta} \right] f(\xi)}, \gamma = 2, 3, \ldots$$

and

$$[\Gamma(.)]^n = \Gamma^n(.).$$

Proof.: Let ξ be a root of f and $\epsilon_i = \xi + e_i$. By Taylor's series expansion of $f(\epsilon_i)$ and $\left[{}_{c}^{\smile}\partial_{\delta_1}^{\delta} \right] f(\epsilon_i)$ around $\epsilon = \xi$, taking $f(\xi) = 0$, we get:

$$f(\epsilon_i) = \frac{\left[{}_{c}^{\smile}\partial_{\xi}^{1\delta} \right] f(\xi)}{\Gamma(\delta+1)} \left[e_i^{\delta} + \breve{c}_2 e_i^{2\delta} + \breve{c}_3 e_i^{3\delta} \right] + O\left(e_i^{4\delta}\right), \tag{17}$$

and

$$\left[{}_{c}^{\smile}\partial_{\delta_1}^{\delta} \right] f(\epsilon_i) = \frac{\left[{}_{c}^{\smile}\partial_{\xi}^{1\delta} \right] f(\xi)}{\Gamma(\delta+1)} \left[\Gamma(\delta+1) + \frac{\Gamma(2\delta+1)}{\Gamma(\delta+1)} \breve{c}_2 e_i^{\delta} + \frac{\Gamma(3\delta+1)}{\Gamma(2\delta+1)} \breve{c}_3 e_i^{2\delta} \right] + O\left(e_i^{3\delta}\right). \tag{18}$$

$$\left(\left[{}_{\breve{c}}\mathcal{D}^{\delta}_{\delta_1}\right]f(\epsilon_i)\right)^{-1} = \frac{1}{\Gamma(\delta+1)} - \frac{\Gamma(2\delta+1)}{\Gamma(\delta+1)}\breve{c}_2 e_i^{\delta} + \frac{\left(-\frac{\Gamma(3\delta+1)}{\Gamma(\delta+1)\Gamma(2\delta+1)} + \frac{(\Gamma(2\delta+1))^2}{(\Gamma(\delta+1))^4}\right)}{\Gamma(\delta+1)}\breve{c}_3 e_i^{2\delta} \tag{19}$$

$$+\left(\frac{\frac{\Gamma(3\delta+1)}{\Gamma(\delta+1)}\breve{c}_2\breve{c}_3 - \frac{(\Gamma(2\delta+1))^3\breve{c}_2^2 - \Gamma(3\delta+1)((\Gamma(\delta+1))^3\breve{c}_2\breve{c}_3}{(\Gamma(\delta+1))^6}}{\Gamma(\delta+1)}\right)e_i^{3\delta} + O\left(e_i^{4\delta}\right). \tag{20}$$

Multiplying (19) with (20), we have:

$$\frac{f(\epsilon_i)}{\left[{}_{\breve{c}}\mathcal{D}^{\delta}_{\delta_1}\right]f(\epsilon_i)} = \frac{1}{\delta\Gamma(\delta)}e_i^{\delta} + \left(\frac{(2^{\delta})^2\Gamma(\delta+\frac{1}{2})\breve{c}_2}{\delta^2(\Gamma((\delta)))^2\sqrt{\pi}} + \frac{\breve{c}_2}{\delta\Gamma((\delta))}\right)e_i^{2\delta} + \tag{21}$$

$$\left(\begin{array}{c}\frac{(2^{\delta})^4(\Gamma(\delta+\frac{1}{2}))^2\breve{c}_2^2}{\delta^3(\Gamma(\delta))^3\pi} - \frac{(2^{\delta})^2\Gamma\left(\delta+\frac{1}{2}\right)\breve{c}_2^3}{\delta^2(\Gamma(\delta))^2\sqrt{\pi}} - \\ \frac{1}{2}\frac{\breve{c}_3\left(3^{\delta}\sqrt{3}\Gamma\left(\delta+\frac{1}{3}\right)\Gamma\left(\delta+\frac{2}{3}\right)\right)}{\delta^2(\Gamma(\delta))^2\sqrt{\pi}(2^{\delta})^2\Gamma\left(\delta+\frac{1}{2}\right)} + \frac{\breve{c}_3}{\delta\Gamma(\delta)}\end{array}\right)e_i^{3\delta} + O\left(e_i^{3\delta}\right), \tag{22}$$

where $\Gamma\left(\frac{1}{\delta}+1\right) = \frac{1}{\delta}\Gamma\left(\frac{1}{\delta}\right)$.

$$\S\left(\frac{f(\epsilon_i)}{\left[{}_{\breve{c}}\mathcal{D}^{\delta}_{\delta_1}\right]f(\epsilon_i)}\right) = \S(0) + \S'(0)\left(\frac{f(\epsilon_i)}{\left[{}_{\breve{c}}\mathcal{D}^{\delta}_{\delta_1}\right]f(\epsilon_i)}\right) + \S''(0)\left(\frac{f(\epsilon_i)}{\left[{}_{\breve{c}}\mathcal{D}^{\delta}_{\delta_1}\right]f(\epsilon_i)}\right)^2 + \ldots \tag{23}$$

$$= \S(0) + \frac{\S'(0)}{\delta\Gamma(\delta)}e_i^{\delta} + \left(\frac{\S''(0)\breve{c}_2}{\delta\Gamma(\delta)} - \frac{\S''(0)(2^{\delta})^2\Gamma((\delta+\frac{1}{2})\breve{c}_2}{\delta^2(\Gamma(\delta))^2\sqrt{\pi}} + \frac{1}{2}\frac{\S''(0)}{\delta^2(\Gamma(\delta))^2}\right)e_i^{2\delta} + \tag{24}$$

$$\left(\begin{array}{c}-\frac{1}{2}\frac{\S'(0)\left(3^{\delta}\sqrt{3}\Gamma\left(\delta+\frac{1}{3}\right)\Gamma\left(\delta+\frac{2}{3}\right)\right)\breve{c}_3}{\delta^2(\Gamma(\delta))^2\sqrt{\pi}(2^{\delta})^2\Gamma\left(\delta+\frac{1}{2}\right)} + \frac{\S'(0)\breve{c}_3}{\delta\Gamma(\delta)} + \frac{\S''(0)\breve{c}_2}{\delta^2(\Gamma(\delta))^2} + \\ \frac{\S'(0)(2^{\delta})^4\,\Gamma^2\left(\delta+\frac{1}{2}\right)\breve{c}_2^2}{\delta^2(\Gamma(\delta))^3\pi} - \frac{\S'(0)(2^{\delta})^2\Gamma\left(\delta+\frac{1}{2}\right)\breve{c}_2^2}{\delta^2(\Gamma(\delta))^2\sqrt{\pi}} - \frac{\S''(0)(2^{\delta})^2\Gamma\left(\delta+\frac{1}{2}\right)\breve{c}_2}{\delta^3(\Gamma(\delta))^3\sqrt{\pi}}\end{array}\right)e_i^{3\delta} + +O\left(e_i^{4\delta}\right) \tag{25}$$

$$\left(\Gamma(\delta+1)\frac{f(\varepsilon_i)}{\left[{}_{\breve{c}}\mathcal{D}^{\delta}_{\delta_1}\right]f(\varepsilon_i)}\right)^{1/\delta} = \S(0)\delta\Gamma(\delta) + \S'(0)e_i + \left(\begin{array}{c}\S'(0)\breve{c}_2 - \frac{\S'(0)(2^{\delta})^2\Gamma\left(\left(\delta+\frac{1}{2}\right)\breve{c}_2\right.}{\delta(\Gamma(\delta))\sqrt{\pi}} \\ +\frac{1}{2}\frac{\S''(0)}{\delta(\Gamma(\delta))}\end{array}\right)e_i^{\delta+1} \tag{26}$$

$$\left(\begin{array}{c} -\dfrac{1}{2}\dfrac{\S'(0)\left(3^{\delta}\sqrt{3}\lceil\left(\delta+\frac{1}{3}\right)\lceil\left(\delta+\frac{2}{3}\right)\right)\breve{c}_3}{\delta(\lceil(\delta))\sqrt{\pi}(2^{\delta})^2\lceil\left(\delta+\frac{1}{2}\right)}+\S'(0)\breve{c}_3+\dfrac{\S''(0)\breve{c}_2}{\delta(\lceil(\delta))}+ \\ \dfrac{\S'(0)(2^{\delta})^4\lceil^2\left(\delta+\frac{1}{2}\right)\breve{c}_2^2}{\delta^2(\lceil(\delta))^2\pi}-\dfrac{\S'(0)(2^{\delta})^2\lceil\left(\delta+\frac{1}{2}\right)\breve{c}_2^2}{\delta(\lceil(\delta))\sqrt{\pi}}-\dfrac{\S''(0)(2^{\delta})^2\lceil\left(\delta+\frac{1}{2}\right)\breve{c}_2}{\delta^2(\lceil(\delta))^2\sqrt{\pi}} \end{array}\right)e_i^{2\delta+1}+O(e_i^{3\delta+1}). \tag{27}$$

$$\begin{aligned} \epsilon_{i+1}=\epsilon_i-&\left(\lceil(\delta+1)\frac{f(\epsilon_i)}{\left[{}_{\breve{c}}\eth_{\delta_1}^{\delta}\right]f(\epsilon_i)+\vartheta^{[*]}f(\epsilon_i)}\right)^{1/\delta}. \\ =-\S(0)\delta\lceil(\delta)+(1-\S'(0))e_i+&\left(\begin{array}{c} -\S'(0)\breve{c}_2-\dfrac{\S'(0)(2^{\delta})^2\lceil\left(\left(\delta+\frac{1}{2}\right)\breve{c}_2\right.}{\delta(\lceil(\delta))\sqrt{\pi}} \\ +\dfrac{1}{2}\dfrac{\S''(0)}{\delta(\lceil(\delta))} \end{array}\right)e_i^{\delta+1} \end{aligned}$$

$$\left(\begin{array}{c} \dfrac{1}{2}\dfrac{\S'(0)\left(3^{\delta}\sqrt{3}\lceil\left(\delta+\frac{1}{3}\right)\lceil\left(\delta+\frac{2}{3}\right)\right)\breve{c}_3}{\delta(\lceil(\delta))\sqrt{\pi}(2^{\delta})^2\lceil\left(\delta+\frac{1}{2}\right)}-\S'(0)\breve{c}_3-\dfrac{\S''(0)\breve{c}_2}{\delta(\lceil(\delta))}+ \\ \dfrac{\S'(0)(2^{\delta})^{42}\left(\delta+\frac{1}{2}\right)\breve{c}_2^2}{\delta^2(\lceil(\delta))^2\pi}+\dfrac{\S'(0)(2^{\delta})^2\lceil\left(\delta+\frac{1}{2}\right)\breve{c}_2^2}{\delta(\lceil(\delta))\sqrt{\pi}}+\dfrac{\S''(0)(2^{\delta})^2\lceil\left(\delta+\frac{1}{2}\right)\breve{c}_2}{\delta^2(\lceil(\delta))^2\sqrt{\pi}} \end{array}\right)e_i^{2\delta+1}++O(e_i^{3\delta+1}). \tag{28}$$

Using the properties $\S(0)=0$ and $\S'(0)=1$ in $e_i^{[*]}$, we have

$$e_i^{[*]}=\left(\frac{(2^{\delta})^2\lceil(\delta+\frac{1}{2})\breve{c}_2}{\delta\lceil(\delta)\sqrt{\pi}}-\breve{c}_2^2-\frac{1}{2}\frac{\S''(0)}{\delta(\lceil(\delta))}\right)e_i^{\delta+1}+O(e_i^{2\delta+1}). \tag{29}$$

Hence, the theorem is proved. □

Remarks: If we choose the weight function $\S\left(\frac{f(\epsilon_i^{[1]})}{f(\epsilon_i)}\right)=\left(\dfrac{\left(\frac{f(\epsilon_i^{[1]})}{f(\epsilon_i)}\right)}{1+\vartheta\left(\frac{f(\epsilon_i^{[1]})}{f(\epsilon_i)}\right)}\right)$ satisfying the conditions $\S(0)=1$, $\S'(0)=0$ and $\S''(0)<\infty$, then we have the following scheme:

$$\epsilon_{i+1}-\xi=\epsilon_i-\xi-\left(\lceil(\delta+1)\frac{\dfrac{f(\epsilon_i)}{\left[{}_{\breve{c}}\eth_{\delta_1}^{\delta}\right]f(\epsilon_i)}}{1+\vartheta^{[*]}\dfrac{f(\epsilon_i)}{\left[{}_{\breve{c}}\eth_{\delta_1}^{\delta}\right]f(\epsilon_i)}}\right)^{1/\delta}, \tag{30}$$

or equivalently,

$$\epsilon_{i+1}=\epsilon_i-\left(\lceil(\delta+1)\frac{f(\epsilon_i)}{\left[{}_{\breve{c}}\eth_{\delta_1}^{\delta}\right]f(\epsilon_i)+\vartheta^{[*]}f(\epsilon_i)}\right)^{1/\delta}. \tag{31}$$

We abbreviate this method as $MM^{[+]}$. The error term for $MM^{[+]}$ is given by:

$$e_i^{[*]} = \left(\frac{(2^\delta)^2 \Gamma(\delta + \frac{1}{2}) \breve{c}_2}{\delta \Gamma(\delta) \sqrt{\pi}} - \breve{c}_2 - \frac{1}{2} \frac{\vartheta^{*]}}{\delta(\Gamma(\delta))} \right) e_i^{\delta+1} + O(e_i^{2\delta+1}). \tag{32}$$

If we choose the weight function $\S\left(\frac{f(\epsilon_i^{[1]})}{f(\epsilon_i)}\right) = \left(\frac{\left(\frac{f(\epsilon_i^{[1]})}{f(\epsilon_i)}\right)}{1-\vartheta\left(\frac{f(\epsilon_i^{[1]})}{f(\epsilon_i)}\right)}\right)$ satisfying the conditions $\S(0) = 1$, $\S'(0) = 0$ and $\S''(0) < \infty$, then we have the following scheme:

$$\epsilon_{i+1} - \xi = \epsilon_i - \xi - \left(\Gamma(\delta - 1) \frac{\frac{f(\epsilon_i)}{\left[{}_{\breve{c}}\partial_{\delta_1}^{\delta}\right] f(\epsilon_i)}}{1 - \vartheta^{[*]} \frac{f(\epsilon_i)}{\left[{}_{\breve{c}}\partial_{\delta_1}^{\delta}\right] f(\epsilon_i)}} \right)^{1/\delta}, \tag{33}$$

or equivalently,

$$\epsilon_{i+1} = \epsilon_i - \left(\Gamma(\delta + 1) \frac{f(\epsilon_i)}{\left[{}_{\breve{c}}\partial_{\delta_1}^{\delta}\right] f(\epsilon_i) - \vartheta^{[*]} f(\epsilon_i)} \right)^{1/\delta}. \tag{34}$$

We abbreviate this method as $MM^{[-]}$. The error term for $MM^{[-]}$ is given by:

$$e_i^{[*]} = \left(-\frac{(2^\delta)^2 \Gamma(\delta + \frac{1}{2}) \breve{c}_2}{\delta \Gamma(\delta) \sqrt{\pi}} + \breve{c}_2 + \frac{1}{2} \frac{\vartheta^{[*]}}{\delta(\Gamma(\delta))} \right) e_i^{\delta+1} + O(e_i^{2\delta+1}). \tag{35}$$

4. Dynamical analysis of the $MM^{[+]}$–$MM^{[-]}$ methods

In this section, we discuss the dynamical analysis of the family of fractional schemes $MM^{[+]}$ and $MM^{[-]}$,

$$\epsilon_{i+1} = \epsilon_i - \left(\Gamma(\delta + 1) \frac{\frac{f(\epsilon_i)}{\left[{}_{\breve{c}}\partial_{\delta_1}^{\delta}\right] f(\epsilon_i)}}{1 \pm \vartheta^{*]} \frac{f(\epsilon_i)}{\left[{}_{\breve{c}}\partial_{\delta_1}^{\delta}\right] f(\epsilon_i)}} \right)^{1/\delta}, \tag{36}$$

for solving (1). Consider the rational map:

$$\epsilon^{[1]} = \epsilon - \left(\Gamma(\delta + 1) \frac{\frac{f(\epsilon_i)}{\left[{}_{\breve{c}}\partial_{\delta_1}^{\delta}\right] f(\epsilon)}}{1 \pm \vartheta^{*]} \frac{f(\epsilon_i)}{\left[{}_{\breve{c}}\partial_{\delta_1}^{\delta}\right] f(\epsilon)}} \right)^{1/\delta}, \tag{37}$$

or, equivalently,

$$\epsilon_{i+1} = \frac{\epsilon\left(\left[{}_{\breve{c}}\partial_{\delta_1}^{\delta}\right] f(\epsilon) \pm \vartheta^{[*]} f(\epsilon_i)\right)^{1/\delta} - \left(\Gamma(\delta+1)\Gamma(\delta+1) f(\epsilon_i)\right)^{1/\delta}}{\left(\left[{}_{\breve{c}}\partial_{\delta_1}^{\delta}\right] f(\epsilon) \pm \vartheta^{[*]} f(\epsilon_i)\right)^{1/\delta}}. \tag{38}$$

We recall some basic concepts of dynamical systems theory (detailed information can be found in Refs. [43, 44]). Consider a rational function $\mathfrak{R}_f : \mathbb{C} \to \mathbb{C}$, where $\mathbb{C}$ denotes the Riemann sphere. The orbit of $\epsilon_0 \in \mathbb{C}$ is defined as the set $orb(\epsilon) = \left\{\epsilon_0, \mathfrak{R}_f(\epsilon_0), \mathfrak{R}_f^2(\epsilon_0), \ldots, \mathfrak{R}_f^m(\epsilon_0), \ldots\right\}$. A point $\epsilon_0 \in \mathbb{C}$ is called a fixed point if $\mathfrak{R}_f(\epsilon_0) = \epsilon_0$. In particular, a fixed point ϵ_0 is called a strange fixed point if $f(\epsilon^r) = 0$ when $\epsilon_0 \neq \epsilon^r$. A T-periodic point is defined as a point $\epsilon_T \in \mathbb{C}$ that satisfies $\mathfrak{R}^T(\epsilon_T) = \epsilon_T$ but $\mathfrak{R}'(\epsilon_T) \neq \epsilon_T$ for $\epsilon < T$. If ϵ_0 is a fixed point of $\mathfrak{R}$, it is classified as follows:

- super attracting if $\left|\mathfrak{R}_f'(\epsilon_0)\right| = 0$,
- attracting if $\left|\mathfrak{R}_f'(\epsilon_0)\right| < 1$,
- repulsive if $\left|\mathfrak{R}_f'(\epsilon_0)\right| > 1$,
- neutral if $\left|\mathfrak{R}_f'(\epsilon_0)\right| = 1$.

A strange fixed point is one that is not associated with any root of the nonlinear Eq. (1). An attracting point $\epsilon^* \in \mathbb{C}$ defines a basin of attraction, $\mathfrak{R}(\epsilon^*)$, which is the set of starting points whose orbit tends to ϵ^*.

To analyze the stability and consistency of the single-step fractional Caputo schemes with respect to the initial guesses used to approximate the root of Eq. (1), we consider the following quadratic polynomials:

$$\mathrm{P}^{[+]}\left(\epsilon^{[*]}\right) = \left(\epsilon^{[*]}\right)^2 + 1, \mathrm{P}^{[+]}\left(\epsilon^{[*]}\right) = \left(\epsilon^{[*]}\right)^2 - 1, \mathrm{P}^{[0]}\left(\epsilon^{[*]}\right) = \left(\epsilon^{[*]}\right)^2. \tag{39}$$

When applying the fractional schemes $\mathrm{MM}^{[+]}$ and $\mathrm{MM}^{[-]}$ to $\mathrm{P}^{[+]}(\epsilon^{[*]})$, $\mathrm{P}^{[-]}(\epsilon^{[*]})$, and $\mathrm{P}^{[0]}(\epsilon^{[*]})$, we obtain the following rational functions. For the iterative scheme $\mathrm{MM}^{[+]}$, the rational functions corresponding to $\mathrm{P}^{[+]}(\epsilon^{[*]})$, $\mathrm{P}^{[-]}(\epsilon^{[*]})$, and $\mathrm{P}^{[0]}(\epsilon^{[*]})$ are given by:

$$\mathfrak{R}^{[\pm]}\left(\epsilon^{[*]}\right) = \epsilon^{[*]} - \left(\frac{\Gamma(\delta+1)\left(\left(\epsilon^{[*]}\right)^2 \pm 1\right)}{2\frac{(\epsilon^{[*]})3-\delta}{\Gamma(3-\delta)} - \frac{(\epsilon^{[*]})-\delta}{\Gamma(1-\delta)} + \vartheta^{[*]}(\epsilon^{[*]})^2 \pm 1}\right)^{\frac{1}{\delta}}, \tag{40}$$

$$\mathfrak{R}^{[0]}\left(\epsilon^{[*]}\right) = \epsilon\varepsilon^{[*]} - \left(\frac{\Gamma(\delta+1)\left(\left(\epsilon^{[*]}\right)^2\right)}{2\frac{(\epsilon^{[*]})3-\delta}{\Gamma(3-\delta)} - \frac{(\epsilon^{[*]})-\delta}{\Gamma(1-\delta)} + \vartheta^{.*]}(\epsilon^{[*]})^2}\right)^{\frac{1}{\delta}}. \tag{41}$$

Similarly, for the iterative scheme $\mathrm{MM}^{[-]}$, the rational maps corresponding to $\mathrm{P}^{[+]}(\epsilon^{[*]})$, $\mathrm{P}^{[-]}(\epsilon^{[*]})$, and $\mathrm{P}^{[0]}(\epsilon^{[*]})$ are:

$$\mathfrak{R}^{[\pm]}\left(\epsilon^{[*]}\right) = \epsilon^{[*]} - \left(\frac{\Gamma(\delta+1)\left(\left(\epsilon^{[*]}\right)^2 \pm 1\right)}{2\frac{(\epsilon^{[*]})3-\delta}{\Gamma(3-\delta)} - \frac{(\epsilon^{[*]})-\delta}{\Gamma(1-\delta)} - \vartheta^{[*]}(\epsilon^{[*]})^2 \pm 1}\right)^{\frac{1}{\delta}}, \tag{42}$$

$$\mathfrak{R}^{[0]}\left(\epsilon^{[*]}\right) = \epsilon^{[*]} - \left(\frac{\Gamma(\delta+1)\left(\left(\epsilon^{[*]}\right)^2\right)}{2\frac{\left(\epsilon^{[*]}\right)^3-\delta}{\Gamma(3-\delta)} - \frac{\left(\epsilon^{[*]}\right)-\delta}{\Gamma(1-\delta)} - \vartheta^{[*]}\left(\epsilon^{[*]}\right)^2}\right)^{\frac{1}{\delta}}. \tag{43}$$

Examining these rational functions provides insight into how initial guesses influence the convergence and dynamical behaviors of the schemes. **Figure 1** shows a flowchart illustrating the stability analysis procedure for fractional-order single-step iterative algorithms. To identify the optimal parameters for robust convergence, the process includes generating parametric planes with critical points, identifying convergence and divergence regions, analyzing dynamical planes for fractal structures and sensitivity, studying fixed and strange fixed points, and assessing local stability.

We now present some theoretical results that analyze the number of fixed points and the corresponding asymptotic performance of the schemes $MM^{[+]}$ and $MM^{[-]}$.

Lemma 1: *The rational operators* $\mathfrak{R}^{[\pm]}\left(\epsilon^{[*]}\right)$ *and* $\mathfrak{R}^{[0]}\left(\epsilon^{[*]}\right)$*, corresponding to the fractional computing schemes* $MM^{[+]}$ *and* $MM^{[-]}$*, coincide with the roots of the associated polynomials. The fixed points are super-attracting for* $\mathfrak{R}^{[\pm]}\left(\epsilon^{[*]}\right)$ *and attracting for* $\mathfrak{R}^{[0]}\left(\epsilon^{[*]}\right)$*. Furthermore, the strange fixed points at* $\epsilon^{[*]} = \infty$ *are always neutral.*

Proof. For the iteration technique $MM^{[+]}$, we solve $\mathfrak{R}^{[\pm]}\left(\epsilon^{[*]}\right) = \epsilon^{[*]}$, which can be equivalently expressed as

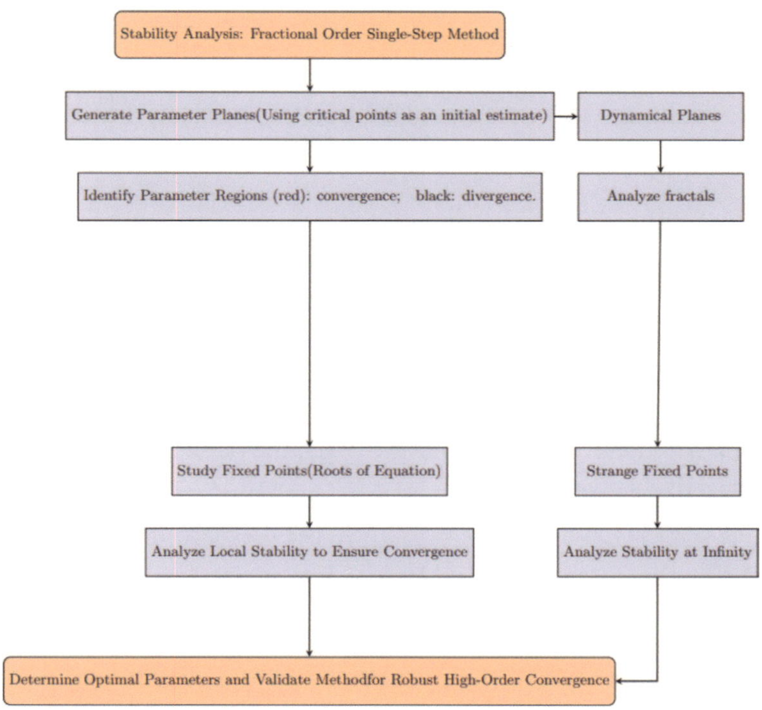

Figure 1.
A schematic illustration of the stability analysis framework used to evaluate fractional-order single-step iterative methods.

$$-\left(\frac{\Gamma(\delta+1)\left(\left(\epsilon^{[*]}\right)^2 \pm 1\right)}{\vartheta^{[*]}\left(\epsilon^{[*]}\right)^2 \pm \vartheta^{[*]} + 2\epsilon^{[*]}}\right)^{\frac{1}{\delta}} = 0, \tag{44}$$

and for $MM^{[-]}$, we have

$$-\left(-\frac{\Gamma(\delta+1)\left(\left(\epsilon^{[*]}\right)^2 \pm 1\right)}{\vartheta^{[*]}\left(\epsilon^{[*]}\right)^2 \pm \vartheta^{[*]} - 2\epsilon^{[*]}}\right)^{\frac{1}{\delta}} = 0. \tag{45}$$

The solutions of (44) and (45) are the roots $\epsilon^{[*]}_{1,[+]} = -1, \epsilon^{[*]}_{2,[+]} = 1, \epsilon^{[*]}_{1,[-]} = i$ and $\epsilon^{[*]}_{2,[-]} = -i$. These fixed points are the roots of the associated polynomials of $\mathfrak{R}^{[\pm]}\left(\epsilon^{[*]}\right)$, for the $MM^{[+]}$ map. Similarly, the fixed points of map $MM^{[-]}$ corresponding to (45) have the same roots. To verify that $\epsilon^{[*]} = \infty$ is a fixed point of the operator $\mathfrak{R}^{[+]}_{\vartheta^{[*]}}\left(\epsilon^{[*]}\right)$ and $\mathfrak{R}^{[-]}_{\vartheta^{[*]}}\left(\epsilon^{[*]}\right)$ for the $MM^{[+]}$ and $MM^{[-]}$ maps, we define

$$\pounds_{\vartheta^{[*]}}\left(\epsilon^{[*]}\right) = \frac{1}{\mathfrak{R}^{[\pm]}_{\vartheta^{[*]}}\left(\frac{1}{\epsilon^{[*]}}\right)}. \tag{46}$$

Using the transformation for the rational operator $\mathfrak{R}^{[\pm]}\left(\epsilon^{[*]}\right)$ corresponding to $MM^{[+]}$, we have

$$\pounds_{\vartheta^{[*]}}\left(\epsilon^{[*]}\right) = -\frac{\epsilon^{[*]}}{\left(\frac{\Gamma(\delta+1)\left(\left(\epsilon^{[*]}\right)^2 \pm 1\right)}{\vartheta^{[*]}\left(\epsilon^{[*]}\right)^2 \pm \vartheta^{[*]} - 2\epsilon^{[*]}}\right)^{\frac{1}{\delta}} \epsilon^{[*]} - 1}, \tag{47}$$

and for $MM^{[-]}$,

$$\pounds_{\vartheta^{[*]}}\left(\epsilon^{[*]}\right) = -\frac{\epsilon^{[*]}}{-\left(\frac{\Gamma(\delta+1)\left(\left(\epsilon^{[*]}\right)^2 \pm 1\right)}{\vartheta^{[*]}\left(\epsilon^{[*]}\right)^2 \pm \vartheta^{[*]} \pm 2\epsilon^{[*]}}\right)^{\frac{1}{\delta}} \epsilon^{[*]} - 1}. \tag{48}$$

Thus, $\pounds_{\vartheta^{[*]}}(0) = 0$ from (47) and (48). So, infinity is a strange fixed point of $MM^{[+]}$ and $MM^{[-]}$ for $\delta \approx 1$. For the polynomial $P^{[0]}\left(\epsilon^{[*]}\right)$, the rational operator of $MM^{[+]}$ and $MM^{[-]}$ is respectively given as:

$$-\left(\frac{\epsilon^{[*]}\Gamma(\delta+1)}{\vartheta^{[*]}\left(\epsilon^{[*]}\right) + 2}\right)^{\frac{1}{\delta}} = \epsilon^{[*]}, \tag{49}$$

and

$$-\left(-\frac{\epsilon^{[*]}\Gamma(\delta+1)}{\vartheta^{[*]}\left(\epsilon^{[*]}\right) + 2}\right)^{\frac{1}{\delta}} = \epsilon^{[*]}. \tag{50}$$

For $\delta \approx 1$, the fixed point is 0, and we denote this root by $\epsilon_0^{[*]}$ for the iterative scheme $\text{MM}^{[+]}$ and $\text{MM}^{[-]}$. For $\mathfrak{R}^{[0]}\left(\epsilon_0^{[*]}\right)$ of $\text{MM}^{[+]}$, we have the following concerning the operator linked to infinity:

$$\pounds_{\vartheta^{[*]}}\left(\epsilon^{[*]}\right) = -\frac{\epsilon^{[*]}}{\left(\frac{\Gamma(\delta+1)}{\vartheta^{[*]}(\epsilon^{[*]})+2}\right)^{\frac{1}{\delta}}\epsilon^{[*]} - 1} \text{corresponding to } MM^{[+]}, \tag{51}$$

$$\pounds_{\vartheta^{[*]}}\left(\epsilon^{[*]}\right) = -\frac{\epsilon^{[*]}}{\left(-\frac{\Gamma(\delta+1)}{\vartheta^{[*]}(\epsilon^{[*]})-2}\right)^{\frac{1}{\delta}}\epsilon^{[*]} - 1} \text{corresponding to } MM^{[-]}. \tag{52}$$

From (51) and (52), as $\pounds_{\vartheta^{[*]}}(0) = 0$, infinity is also a strange fixed point for $\text{P}^{[0]}(\epsilon^{[*]})$ of the single-step fractional scheme $\text{MM}^{[+]}$ and $\text{MM}^{[-]}$. To analyze the local behavior of the fixed points for the single-step methods $\text{MM}^{[+]}$ and $\text{MM}^{[-]}$, we compute the derivative of their associated rational functions, that is, $\left(\mathfrak{R}^{[\pm]}(\epsilon^{[*]})\right)'$. In particular, for the method $\text{MM}^{[+]}$, corresponding to the rational function $P^{[+]}(\epsilon^{[*]})$, we obtain: For $\text{MM}^{[+]}$ of $\text{P}^{[+]}(\epsilon^{[*]})$, we have

$$\left(\mathfrak{R}^{[\pm]}\left(\epsilon^{[*]}\right)\right)' = \frac{\begin{pmatrix} \pm\vartheta^{[*]}\delta(\epsilon^{[*]})^4 + 2\left(\frac{\Gamma(\delta+1)\left((\epsilon^{[*]})^2 \pm 1\right)}{\vartheta^{[*]}(\epsilon^{[*]})^2 \pm \vartheta^{[*]} \pm 2\epsilon^{[*]}}\right)(\epsilon^{[*]})^4 + \\ 2\vartheta^{[*]}\delta(\epsilon^{[*]})^2 + 2\delta(\epsilon^{[*]})^3 \pm \\ 2\left(\frac{\Gamma(\delta+1)\left((\epsilon^{[*]})^2 \pm 1\right)}{\vartheta^{[*]}(\epsilon^{[*]})^2 \pm \vartheta^{[*]} \pm 2\epsilon^{[*]}}\right)(\epsilon^{[*]})^2 \pm \vartheta^{[*]}\delta \pm 2\vartheta^{[*]}\delta(\epsilon^{[*]}) \end{pmatrix}}{\delta(\epsilon^{[*]})^2\left(\vartheta^{[*]}(\epsilon^{[*]})^2 \pm \vartheta^{[*]} \pm 2\epsilon^{[*]}\right)\left((\epsilon^{[*]})^2 \pm 1\right)}, \tag{53}$$

and for $\text{P}^{[0]}(\epsilon^{[*]})$,

$$\left(\mathfrak{R}^{[0]}\left(\epsilon^{[*]}\right)\right)' = \frac{2\left(\frac{\Gamma(\delta+1)}{\vartheta^{[*]}+2\epsilon^{[*]}}\right)(\epsilon^{[*]})^2 - \vartheta^{[*]}\delta - 2\delta\epsilon^{[*]}}{\delta\left(\vartheta^{[*]} - 2\epsilon^{[*]}\right)(\epsilon^{[*]})^2}, \tag{54}$$

and for the iterative scheme,

$$\left(\mathfrak{R}^{[\pm]}\left(\epsilon^{[*]}\right)\right)' = \frac{\begin{pmatrix} \pm\vartheta^{[*]}\delta(\epsilon^{[*]})^4 + 2\left(-\frac{\Gamma(\delta+1)\left((\epsilon^{[*]})^2 \pm 1\right)}{\vartheta^{[*]}(\epsilon^{[*]})^2 \pm \vartheta^{[*]} \pm 2\epsilon^{[*]}}\right)(\epsilon^{[*]})^4 - \\ 2\vartheta^{[*]}\delta(\epsilon^{[*]})^2 + 2\delta(\epsilon^{[*]})^3 \pm \\ 2\left(\frac{\Gamma(\delta+1)\left((\epsilon^{[*]})^2 \pm 1\right)}{\vartheta^{[*]}(\epsilon^{[*]})^2 \pm \vartheta^{[*]} \pm 2\epsilon^{[*]}}\right)(\epsilon^{[*]})^2 \pm \vartheta^{[*]}\delta \pm 2\vartheta^{[*]}\delta(\epsilon^{[*]}) \end{pmatrix}}{\delta(\epsilon^{[*]})^2\left(\vartheta^{[*]}(\epsilon^{[*]})^2 \pm \vartheta^{[*]} - 2\epsilon^{[*]}\right)\left((\epsilon^{[*]})^2 \pm 1\right)}, \tag{55}$$

and for $P^{[0]}(\epsilon^{[*]})$,

$$\left(\mathfrak{R}^{[0]}\left(\epsilon^{[*]}\right)\right)' = \frac{2\left(\frac{\Gamma(\delta+1)}{\vartheta^{[*]}-2\epsilon^{[*]}}\right)\left(\epsilon^{[*]}\right)^2 + \vartheta^{[*]}\delta - 2\delta\epsilon^{[*]}}{\delta\left(\vartheta^{[*]} - 2\epsilon^{[*]}\right)\left(\epsilon^{[*]}\right)^2}. \tag{56}$$

For $\delta \approx 1$, it is easy to prove from (55) and (56) that they contain the term $\left(\epsilon^{[*]}\right)^2 \pm 1$ and $\epsilon_{j,[+]}^{[*]}, \epsilon_{j,[-]}^{[*]}$ for $j = 1, 2$ are super-attracting points of $MM^{[+]}$ and $MM^{[-]}$, respectively, while $\epsilon_0^{[*]}$ is an attracting point because $\left(\mathfrak{R}^{[\pm]}\left(\epsilon^{[*]}\right)\right)' \leq 1$. To assess the behavior at infinity, we evaluate the derivative $\left(\pounds_{\vartheta^{[*]}}\left(\epsilon^{[*]}\right)\right)'$ corresponding to the polynomial mappings $P^{[\pm]}\left(\epsilon^{[*]}\right)$ and $P^{[0]}\left(\epsilon^{[*]}\right)$. For $MM^{[+]}$, we have

$$\left(\pounds_{\vartheta^{[*]}}\left(\epsilon^{[*]}\right)\right)' = \frac{\left(\begin{array}{c} \pm\vartheta^{[*]}\delta\left(\epsilon^{[*]}\right)^4 + 2\left(\frac{\Gamma(\delta+1)\left(\left(\epsilon^{[*]}\right)^2 \pm 1\right)}{\vartheta^{[*]}\left(\epsilon^{[*]}\right)^2 \pm \vartheta^{[*]} \pm 2\epsilon^{[*]}}\right)^{\frac{1}{\delta}}\left(\epsilon^{[*]}\right)^4 + \\ 2\vartheta^{[*]}\delta\left(\epsilon^{[*]}\right)^2 + 2\delta\left(\epsilon^{[*]}\right)^3 \pm \\ 2\left(\frac{\Gamma(\delta+1)\left(\left(\epsilon^{[*]}\right)^2 \pm 1\right)}{\vartheta^{[*]}\left(\epsilon^{[*]}\right)^2 \pm \vartheta^{[*]} \pm 2\epsilon^{[*]}}\right)^{\frac{1}{\delta}}\left(\epsilon^{[*]}\right)^2 \pm \vartheta^{[*]}\delta \pm 2\vartheta^{[*]}\delta\left(\epsilon^{[*]}\right) \end{array}\right)}{\delta\left(\epsilon^{[*]}\right)^2\left(\vartheta^{[*]}\left(\epsilon^{[*]}\right)^2 \pm \vartheta^{[*]} \pm 2\epsilon^{[*]}\right)\left(\left(\epsilon^{[*]}\right)^2 \pm 1\right)}, \tag{57}$$

and for $P^{[0]}(\epsilon^{[*]})$,

$$\left(\pounds_{\vartheta^{[*]}}\left(\epsilon^{[*]}\right)\right)' = \frac{2\left(\frac{\Gamma(\delta+1)}{\vartheta^{[*]}+2\epsilon^{[*]}}\right)^{\frac{1}{\delta}}\left(\epsilon^{[*]}\right)^2 - \vartheta^{[*]}\delta - 2\delta\epsilon^{[*]}}{\left(\left(\frac{\Gamma(\delta+1)}{\vartheta^{[*]}+2\epsilon^{[*]}}\right)^{\frac{1}{\delta}}\epsilon^{[*]} - 1\right)^2\delta\left(2\epsilon^{[*]} + \vartheta^{[*]}\right)}, \tag{58}$$

and for the map,

$$\left(\pounds_{\vartheta^{[*]}}\left(\epsilon^{[*]}\right)\right)' = \frac{\left(\begin{array}{c} \pm\vartheta^{[*]}\delta\left(\epsilon^{[*]}\right)^4 + 2\left(\frac{\Gamma(\delta+1)\left(\left(\epsilon^{[*]}\right)^2 \pm 1\right)}{\vartheta^{[*]}\left(\epsilon^{[*]}\right)^2 \pm \vartheta^{[*]} \pm 2\epsilon^{[*]}}\right)^{\frac{1}{\delta}}\left(\epsilon^{[*]}\right)^4 + \\ 2\vartheta^{[*]}\delta\left(\epsilon^{[*]}\right)^2 + 2\delta\left(\epsilon^{[*]}\right)^3 \pm \\ 2\left(\frac{\Gamma(\delta+1)\left(\left(\epsilon^{[*]}\right)^2 \pm 1\right)}{\vartheta^{[*]}\left(\epsilon^{[*]}\right)^2 \pm \vartheta^{[*]} \pm 2\epsilon^{[*]}}\right)^{\frac{1}{\delta}}\left(\epsilon^{[*]}\right)^2 \pm \vartheta^{[*]}\delta \pm 2\vartheta^{[*]}\delta\left(\epsilon^{[*]}\right) \end{array}\right)}{\delta\left(\epsilon^{[*]}\right)^2\left(\vartheta^{[*]}\left(\epsilon^{[*]}\right)^2 \pm \vartheta^{[*]} \pm 2\epsilon^{[*]}\right)\left(\left(\epsilon^{[*]}\right)^2 \pm 1\right)}, \tag{59}$$

and for $P^{[0]}(\epsilon^{[*]})$,

$$\left(\pounds_{\vartheta^{[*]}}\left(\epsilon^{[*]}\right)\right)' = \frac{2\left(\frac{\Gamma(\delta+1)}{\vartheta^{[*]}+2\epsilon^{[*]}}\right)^{\frac{1}{\delta}}\left(\epsilon^{[*]}\right)^2 - \vartheta^{[*]}\delta - 2\delta\epsilon^{[*]}}{\left(\left(-\frac{\Gamma(\delta+1)}{\vartheta^{[*]}-2\epsilon^{[*]}}\right)^{\frac{1}{\delta}}\epsilon^{[*]} - 1\right)^2\delta\left(-2\epsilon^{[*]} + \vartheta^{[*]}\right)}. \tag{60}$$

Therefore, for $\delta \approx 1$ and $\epsilon^{[*]} = 0$, we find that $\left(\pounds_{\vartheta^{[*]}}(0)\right)' = 1$. Thus, infinity serves as a neutral point in both cases. This completes the proof. □

The following lemma 2 provides valuable insights into the critical points of the rational operator $\mathfrak{R}^{[\pm]}(\epsilon^{[*]}), \mathfrak{R}^{[0]}(\epsilon^{[*]})$ of the fractional iterative scheme.

Lemma 2:

1.For $\mathfrak{R}^{[-]}(\epsilon^{[*]})$, the fixed points of the $MM^{[+]}$ are critical points. Specifically, when $\vartheta^{[*]} \neq \pm i$ and $\vartheta^{[*]} \neq 0$, then $\mathfrak{R}^{[-]}(\epsilon^{[*]})$ of $MM^{[+]}$ has two free critical points, namely

$$\mathtt{C}_{11}^{[r]} = \frac{-2 - \sqrt{2 + (\vartheta^{[*]})^2}}{\vartheta^{[*]}}, \mathtt{C}_{12}^{[r]} = \frac{-2 + \sqrt{2 + (\vartheta^{[*]})^2}}{\vartheta^{[*]}}. \tag{61}$$

When $\vartheta^{[*]} = \pm i$, it has no free critical points. Similarly, for the fractional Caputo scheme $MM^{[-]}$, we have

$$\mathtt{C}_{13}^{[r]} = \frac{2 + \sqrt{2 + (\vartheta^{[*]})^2}}{\vartheta^{[*]}}, \mathtt{C}_{14}^{[r]} = \frac{2 - \sqrt{2 + (\vartheta^{[*]})^2}}{\vartheta^{[*]}}. \tag{62}$$

2.For $\mathfrak{R}^{[+]}(\epsilon^{[*]})$, the fixed points of $MM^{[+]}$ are critical points. Specifically, when $\vartheta^{[*]} \neq \pm 1$ and $\vartheta^{[*]} = 0$, then $\mathfrak{R}^{[+]}(\epsilon^{[*]})$ has two free critical points, namely

$$\mathtt{C}_{21}^{[r]} = \frac{-2 - \sqrt{2 - (\vartheta^{[*]})^2}}{\vartheta^{[*]}}, \mathtt{C}_{22}^{[r]} = \frac{-2 + \sqrt{2 - (\vartheta^{[*]})^2}}{\vartheta^{[*]}}. \tag{63}$$

When $\vartheta^{[*]} = \pm 1$, it has no free critical points. Similarly, for the fractional Caputo scheme $MM^{[-]}$, we have

$$\mathtt{C}_{23}^{[r]} = \frac{2 + \sqrt{2 - (\vartheta^{[*]})^2}}{\vartheta^{[*]}}, \mathtt{C}_{24}^{[r]} = \frac{2 - \sqrt{2 - (\vartheta^{[*]})^2}}{\vartheta^{[*]}}. \tag{64}$$

3.If $\vartheta^{[*]} \neq 0$, the operator $\mathfrak{R}^{[0]}(\epsilon^{[*]})$ of $MM^{[+]}$ has two free critical points, namely

$$\mathtt{C}_{31}^{[r]} = \frac{-2 - \sqrt{2}}{\vartheta^{[*]}}, \mathtt{C}_{32}^{[r]} = \frac{-2 + \sqrt{2}}{\vartheta^{[*]}}. \tag{65}$$

Similarly, for the fractional Caputo scheme $MM^{[-]}$, we have

$$\mathtt{C}_{33}^{[r]} = \frac{2 + \sqrt{2}}{\vartheta^{[*]}}, \mathtt{C}_{34}^{[r]} = \frac{2 - \sqrt{2}}{\vartheta^{[*]}}. \tag{66}$$

Proof. To determine the critical points of $MM^{[+]}$ and $MM^{[-]}$ for the polynomials $P^{[-]}(\epsilon^{[*]})$, we set $\left(\mathfrak{R}^{[-]}(\epsilon^{[*]})\right)'$ and $\left(\mathfrak{R}^{[+]}(\epsilon^{[*]})\right)'$ to zero, that is,

$$\frac{\begin{pmatrix} -\vartheta^{[*]}\delta\left(\epsilon^{[*]}\right)^4+2\left(\dfrac{\Gamma(\delta+1)\left(\left(\epsilon^{[*]}\right)^2-1\right)}{\vartheta^{[*]}\left(\epsilon^{[*]}\right)^2-\vartheta^{[*]}-2\epsilon^{[*]}}\right)\left(\epsilon^{[*]}\right)^4+ \\ 2\vartheta^{[*]}\delta\left(\epsilon^{[*]}\right)^2+2\delta\left(\epsilon^{[*]}\right)^3- \\ 2\left(\dfrac{\Gamma(\delta+1)\left(\left(\epsilon^{[*]}\right)^2-1\right)}{\vartheta^{[*]}\left(\epsilon^{[*]}\right)^2-\vartheta^{[*]}-2\epsilon^{[*]}}\right)\left(\epsilon^{[*]}\right)^2-\vartheta^{[*]}\delta-2\vartheta^{[*]}\delta\left(\epsilon^{[*]}\right) \end{pmatrix}}{\delta\left(\epsilon^{[*]}\right)^2\left(\vartheta^{[*]}\left(\epsilon^{[*]}\right)^2-\vartheta^{[*]}-2\epsilon^{[*]}\right)\left(\left(\epsilon^{[*]}\right)^2-1\right)}=0, \tag{67}$$

and

$$\frac{\begin{pmatrix} -\vartheta^{[*]}\delta\left(\epsilon^{[*]}\right)^4+2\left(-\dfrac{\Gamma(\delta+1)\left(\left(\epsilon^{[*]}\right)^2-1\right)}{\vartheta^{[*]}\left(\epsilon^{[*]}\right)^2-\vartheta^{[*]}-2\epsilon^{[*]}}\right)\left(\epsilon^{[*]}\right)^4- \\ 2\vartheta^{[*]}\delta\left(\epsilon^{[*]}\right)^2+2\delta\left(\epsilon^{[*]}\right)^3- \\ 2\left(\dfrac{\Gamma(\delta+1)\left(\left(\epsilon^{[*]}\right)^2-1\right)}{\vartheta^{[*]}\left(\epsilon^{[*]}\right)^2-\vartheta^{[*]}-2\epsilon^{[*]}}\right)\left(\epsilon^{[*]}\right)^2-\vartheta^{[*]}\delta-2\vartheta^{[*]}\delta\left(\epsilon^{[*]}\right) \end{pmatrix}}{\delta\left(\epsilon^{[*]}\right)^2\left(\vartheta^{[*]}\left(\epsilon^{[*]}\right)^2-\vartheta^{[*]}-2\epsilon^{[*]}\right)\left(\left(\epsilon^{[*]}\right)^2-1\right)}=0 \tag{68}$$

Solving (67) and (68) for $\delta \approx 1$ yields the critical points of $MM^{[+]}$ as:

$$C_{11}^{[r]}=\frac{-2-\sqrt{2+\left(\vartheta^{[*]}\right)^2}}{\vartheta^{[*]}},\ C_{12}^{[r]}=\frac{-2+\sqrt{2+\left(\vartheta^{[*]}\right)^2}}{\vartheta^{[*]}},$$

± 1 and for $MM^{[-]}$

$$C_{13}^{[r]}=\frac{2+\sqrt{2+\left(\vartheta^{[*]}\right)^2}}{\vartheta^{[*]}},\ C_{14}^{[r]}=\frac{2-\sqrt{2+\left(\vartheta^{[*]}\right)^2}}{\vartheta^{[*]}},$$

and $\pm i$. Similarly, we compute the remaining critical points ($C_{21}^{[r]}-C_{24}^{[r]}, C_{31}^{[r]}-C_{34}^{[r]}$) of $MM^{[+]}$, $MM^{[-]}$ for polynomials $P^{[+]}\left(\epsilon^{[*]}\right)$, and $P^{[0]}\left(\epsilon^{[*]}\right)$, respectively. □

4.1 Parametric planes of $MM^{[+]}$–$MM^{[-]}$

Parametric planes play a pivotal role in understanding iterative methods devised for solving nonlinear equations. These planes offer a comprehensive view of the methods' behavior and convergence characteristics. Essentially, they encompass all points representing diverse combinations of parameters that define the iterative process. These parameters include initial guesses, convergence criteria, and iterative functions, among others. By plotting iterative trajectories on this plane, one gains insight into patterns related to stability, divergence, and convergence. Iterative methods are designed to iteratively refine initial estimates until they converge to the roots of nonlinear equations. The parametric plane serves as a visual aid, shedding light on the method's sensitivity to various parameter selections. This understanding is instrumental in choosing appropriate parameters that facilitate efficient convergence.

To generate a parametric plane, we apply $\vartheta^{[*]}$ to a mesh of 250,000 points on the complex plane. The iteration process is initialized using critical points as initial guesses and proceeds until convergence to a fixed point is achieved. The complex value of the parameter in the plane is then recorded. If the parameter value converges, the corresponding point is colored red; otherwise, it is colored black.

Utilizing Lemma 2, we focus on the free critical points $\mathtt{C}_{11}^{[r]} - \mathtt{C}_{12}^{[r]}$ and $\mathtt{C}_{13}^{[r]} - \mathtt{C}_{14}^{[r]}$ of the $MM^{[+]}$ and $MM^{[-]}$ methods for each complex value. Points where the method converges to zero or infinity are colored red in the complex plane; all other points are marked black to indicate divergence. **Figures 2(a–c)** and **3(a–c)** were generated for $\vartheta^{[*]}$ values in the range $[-10, 10] \times [-10, 10]$, using a mesh of 250,000 points and performing 20 iterations per point. As shown in **Figures 2** and **3**, the red regions indicate higher stability of the scheme for the corresponding parameter values, while the black regions indicate instability or divergence. Selecting parameter values from the red regions results in a higher convergence rate.

Figures 2(a–c) and **3(a–c)** show parametric planes generated by using the critical points $\mathtt{C}_{11}^{[r]} - \mathtt{C}_{12}^{[r]}$ and $\mathtt{C}_{13}^{[r]} - \mathtt{C}_{14}^{[r]}$ as initial guesses in the fractional iterative schemes $MM^{[+]}$ and $MM^{[-]}$, respectively. The parameter $\vartheta^{[*]}$ determines the color of each pixel based on whether the iteration converges to the root (red) or diverges (black). These planes visually depict the convergence regions and parameter sensitivity of the schemes, providing insight into their robustness to fractional perturbations. By selecting $\vartheta^{[*]}$ values from the red regions of the parametric planes, the approach achieves higher convergence rates and maintains stable behavior. In contrast, parameters chosen outside these regions frequently result in divergence, as demonstrated by the dynamical planes in the following subsections.

4.2 Dynamical planes of $MM^{[+]}$–$MM^{[-]}$

The dynamical plane is essential for understanding the behavior of iterative techniques for solving nonlinear equations. It provides a geometric framework for depicting the iteration process. The iterations of the method generate a sequence of points that may converge to a root, with each point on the plane representing a complex number that denotes the initial approximation or the root estimate. The movement of these

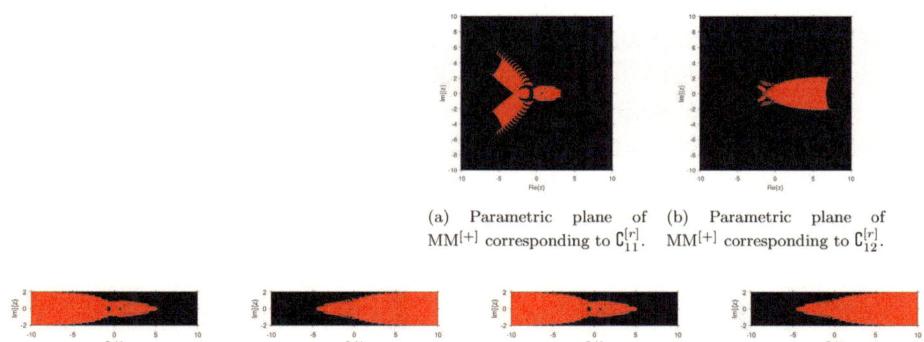

(a) Parametric plane of $MM^{[+]}$ corresponding to $\mathtt{C}_{11}^{[r]}$.

(b) Parametric plane of $MM^{[+]}$ corresponding to $\mathtt{C}_{12}^{[r]}$.

(c) Details of the parametric planes of $MM^{[+]}$ corresponding to critical points $\mathtt{C}_{11}^{[r]}$ and $\mathtt{C}_{12}^{[r]}$, from left to right, indicating the boundaries of the convergence region.

Figure 2.
(a–c): Parametric planes of the single-step fractional schemes $MM^{[+]}$ used to construct the convergence-divergence region of the scheme corresponding to the parameter $\vartheta^{[*]}$, employing the critical points $\mathtt{C}_{11}^{[r]}$ and $\mathtt{C}_{12}^{[r]}$.

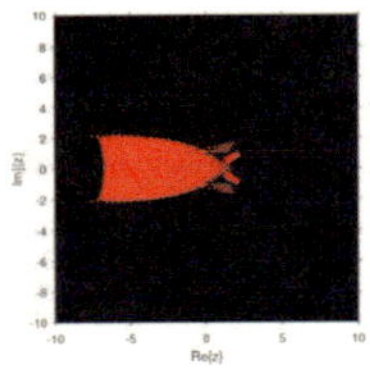

(a) Parametric plane of $MM^{[-]}$ corresponding to $C_{13}^{[r]}$

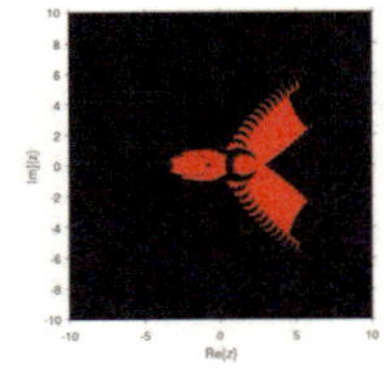

(b) Parametric plane of $MM^{[-]}$ corresponding to $C_{14}^{[r]}$

(c) Details of the parametric planes of $MM^{[-]}$ corresponding to critical points $C_{13}^{[r]}$ and $C_{14}^{[r]}$, from left to right, indicating the boundaries of the convergence region.

Figure 3.
(a–c): Parametric planes of the single-step fractional schemes $MM^{[-]}$ *used to construct the convergence-divergence region of the scheme corresponding to the parameter* $\vartheta^{[*]}$, *employing the critical points* $C_{13}^{[r]}$ *and* $C_{14}^{[r]}$.

points on the plane reveals the dynamics of the iterations, showcasing patterns such as convergence, divergence, periodicity, or chaotic behavior. The dynamical plane provides useful insights into the method's convergence behavior, allowing for the identification of appropriate initial guesses and parameter settings that improve convergence toward the exact solution. Similar to parametric planes, dynamical planes are constructed. Using the same stopping conditions as in the parametric plane, the real and imaginary parts of the initial estimates are represented as two axes on a 250×250 mesh. The dynamical planes of the fractional Caputo-type schemes $MM^{[+]}$ and $MM^{[-]}$ also depend on the parameter value used in the iteration process. Various colors are designated to represent the roots to which the approach converges: orange for points converging to zero or infinity, blue for points converging to one, and black for regions where the points do not converge to any root. The methods $MM^{[+]}$ and $MM^{[-]}$ exhibit stable behavior when parameter values are chosen from the red region of the parametric plane, as shown in **Figure 4(a–e)** for scheme $MM^{[+]}$ and **Figure 5(a–e)** for scheme $MM^{[-]}$. Conversely, they exhibit unstable behavior when parameter values are chosen from the black region of the parametric plane, as shown in **Figures 6(a,b)** and **7(a,b)**, respectively. All dynamical and parametric planes are constructed for $\delta = 0.7$, but similar behavior is observed for other values of δ.

The dynamical planes for the fractional Caputo-type scheme $MM^{[+]}$ with various parameter values $\vartheta^{[*]}$ are illustrated in **Figure 4(a–e)**. This figure represents $\vartheta^{[*]}$ values of 0.3, $-0.4 + 2.3i$, $-0.6 - 0.4i$,$1.5 + 0.03i$,$0.1i$. Notably, when $\vartheta^{*]}$ is 0.3 and $-0.6 - 0.4i$, the $MM^{[+]}$ iteration scheme converges only to zero or infinity, or diverges, as indicated by the orange and black regions. **Figure 4(a,c)** displays a larger orange zone compared to **Figure 4(b,d,e)**, indicating enhanced stability of the $MM^{[+]}$ scheme for these parameter values. Therefore, for the $MM^{[+]}$ iteration scheme, we opted for a parameter value of 0.3, enhancing stability and providing a faster convergence region when solving (1).

The dynamical planes of the fractional Caputo-type scheme $MM^{[-]}$ for various parameter values $\vartheta^{[*]}$ are shown in **Figure 5(a–e)**. These planes correspond to $\vartheta^{[*]}$ values of $1.0 - 0.001i$, $-0.5 - 1.0i$, 0.3, 2.0, and $0.5i$. Notably, when $\vartheta^{[*]}$ is $1.0 - 0.001i$, 0.3, or $0.5i$, the $MM^{[-]}$ iteration scheme converges only to zero, infinity,

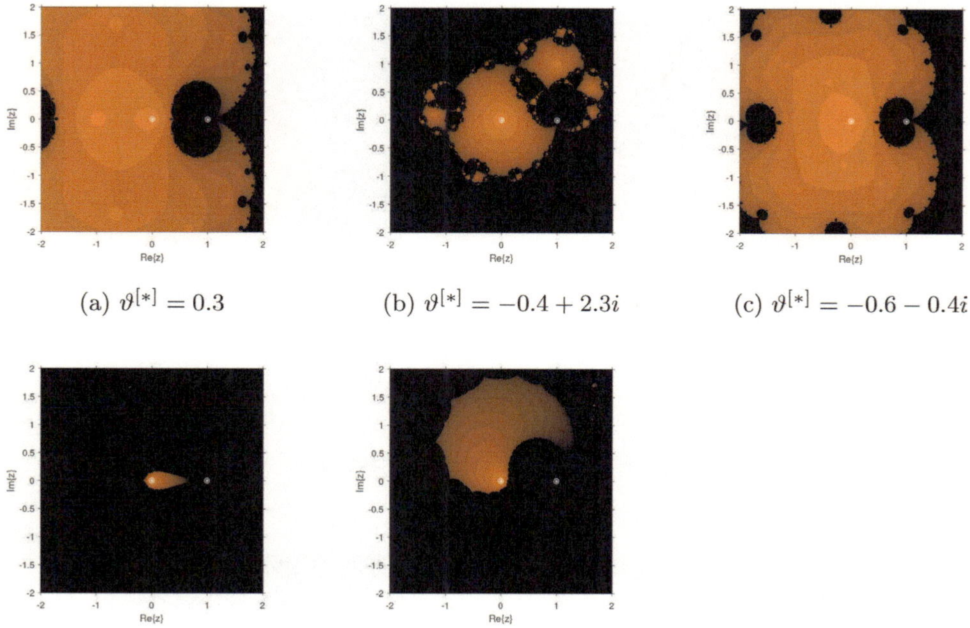

(a) $\vartheta^{[*]} = 0.3$ (b) $\vartheta^{[*]} = -0.4 + 2.3i$ (c) $\vartheta^{[*]} = -0.6 - 0.4i$

(d) $\vartheta^{[*]} = 1.5 + 0.03i$ (e) $\vartheta^{[*]} = 0.1i$

Figure 4.
(a–e): Dynamical planes of the fractional scheme $MM^{[+]}$ *for various parameter values* $\vartheta^{[*]}$.

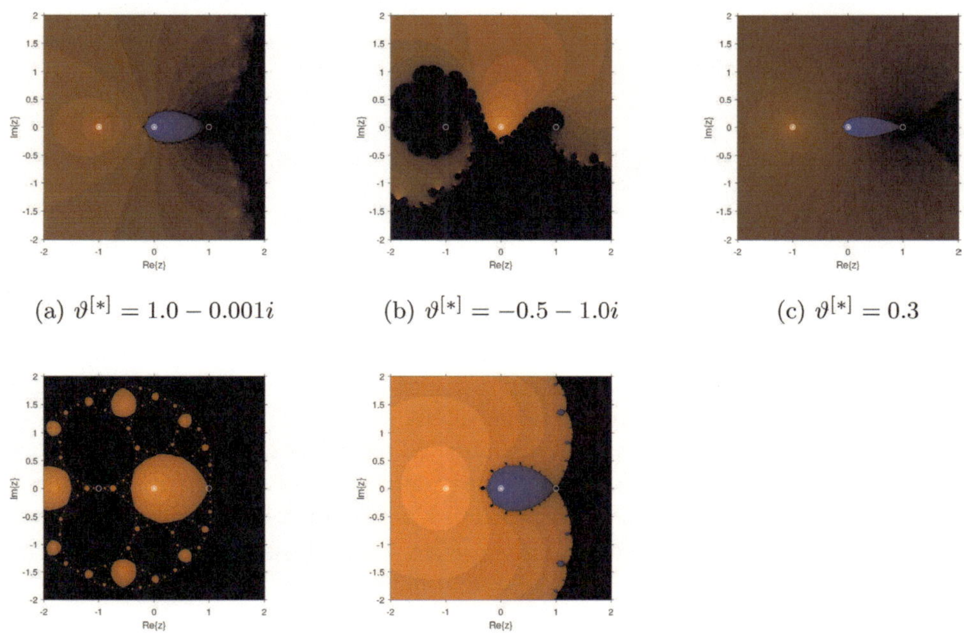

(a) $\vartheta^{[*]} = 1.0 - 0.001i$ (b) $\vartheta^{[*]} = -0.5 - 1.0i$ (c) $\vartheta^{[*]} = 0.3$

(d) $\vartheta^{[*]} = 2.0$ (e) $\vartheta^{[*]} = 0.5i$

Figure 5.
(a–e): Dynamical planes of the fractional scheme $MM^{[-]}$ *for various parameter values* $\vartheta^{[*]}$.

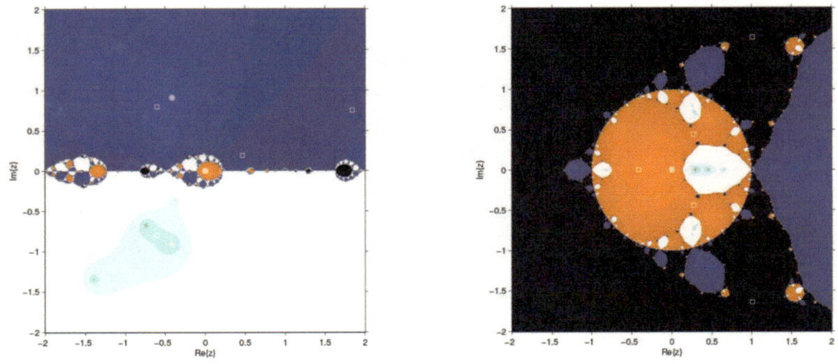

(a) $\vartheta^{[*]} = 150 + 105i$ (b) $\vartheta^{[*]} = -5 - 15i$

Figure 6.
(a–b): Unstable behavior of the scheme $MM^{[+]}$ for parameter $\vartheta^{[*]}$ values selected from the black region of the parametric plane.

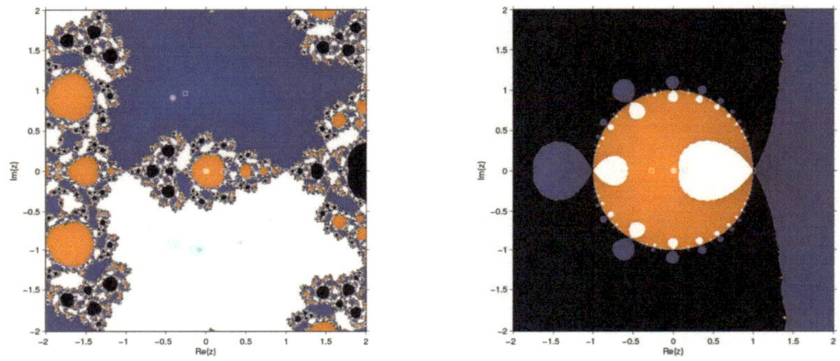

(a) $\vartheta^{[*]} = 15 + 15i$ (b) $\vartheta^{[*]} = 15 - 15i$

Figure 7.
(a–b): Unstable behavior of the scheme $MM^{[-]}$ for parameter $\vartheta^{[*]}$ values selected from the black region of the parametric plane.

or diverges, as indicated by the orange and black regions. In comparison, **Figure 5(a), (c), (e)** exhibit larger orange zones than **Figure 5(b), (d)**, signifying increased stability of the $MM^{[-]}$ scheme for these parameter values. Therefore, for the $MM^{[-]}$ iteration scheme, we selected a parameter value of 0.3, which enhances stability and provides a faster convergence region when solving (1).

Figures 2–7 present the dynamical analysis, which offers important insights into the stability and convergence behavior of the proposed fractional iterative schemes. In **Figures 2** and **3**, parametric planes are used to identify suitable ranges for the control parameter $\vartheta^{[*]}$. Red regions indicate convergence, while black regions indicate divergence. The dynamical planes (**Figures 4** and **5**) illustrate the basins of attraction, whose structure varies with $\vartheta^{[*]}$. Smooth and connected basins in the red regions demonstrate significant convergence and stability, whereas fragmented or fractal-like boundaries near the black zones indicate sensitivity and potential instability. This phenomenon is further demonstrated in **Figures 6** and **7**, where divergent iteration patterns result from parameter values selected from unstable zones. Overall, the dynamical visualizations not only support the theoretical analysis but also serve as a practical reference for parameter adjustment and scheme selection.

5. Numerical results

In this section, we conduct several numerical experiments to illustrate the computational performance of our methods using various engineering applications. The terminating criteria for the computer algorithm implemented in Maple 18 are defined as follows:

$$(i) \qquad e_i = |\epsilon_{i+1} - \epsilon_i| = \Lambda^{[*]} < \in = 10^{-6}, (ii) \qquad e_i = |f(\epsilon_{i+1})| = \Lambda^{[**]} < \in = 10^{-6},$$

where e_i represents the error in the iterative scheme. We compare our family of methods with the Caputo-type fractional version of the classical Newton's method. Candelario et al. [45] presented the following variant:

$$\epsilon_{i+1} = \epsilon_i - \left(\Gamma(\delta+1) \frac{f(\epsilon_i)}{\left[{}_{c}^{\smile}\partial_{\delta_1}^{\delta}\right] f(\epsilon_i)} \right)^{1/\delta}, \tag{69}$$

where $\left[{}_{c}^{\smile}\partial_{\delta_1}^{\delta}\right] f(\epsilon_i) \approx \left[{}_{c}^{\smile}\partial_{\xi}^{\delta}\right] f(\xi)$ for any $\delta \in \mathbb{R}$. The fractional Newton method (abbreviated as $MM^{[*]}$) exhibits an order of convergence of $\delta + 1$ and satisfies the following error equation:

$$e_i^{[*]} = \frac{\Gamma(\delta+1) - \Gamma(\delta+1)}{\Gamma(\delta+1)} \breve{c}_2 e_i^{\delta+1} + O\left(e_i^{2\delta+1}\right), \tag{70}$$

where $e_i^{[*]} = \epsilon_{i+1} - \xi$ and $e_i = \epsilon_i - \xi$, and $\breve{c}_j = \frac{\Gamma(\delta+1)}{\Gamma(\gamma\delta+1)} \frac{\left[{}_{c}^{\smile}\partial_{\xi}^{\gamma\delta}\right] f(\xi)}{\left[{}_{c}^{\smile}\partial_{\xi}^{\delta}\right] f(\xi)}, \gamma = 2, 3, \ldots$.

Shams et al. [46] proposed the following single-step fractional iterative method:

$$\epsilon_{i+1} = \epsilon_i - \left(\Gamma(\delta+1) \frac{f(\epsilon_i)}{\left[{}_{c}^{\smile}\partial_{\delta_1}^{\delta}\right] f(\epsilon_i)} \left[\frac{1}{1 - \alpha \frac{f(\epsilon_i)}{1+f(\epsilon_i)}} \right] \right)^{1/\delta}, \tag{71}$$

The order of convergence of this technique, denoted as $MM^{[**]}$), is $\delta + 1$ and satisfies the following error equation.

$$e_i^{[*]} = \frac{(\alpha + \breve{c}_2)\Gamma^2(\delta+1) - \breve{c}_2 \Gamma(2\delta+1)}{\delta \Gamma(\delta+1)} \breve{c}_2 e_i^{\delta+1} + O\left(e_i^{2\delta+1}\right), \tag{72}$$

where $e_i^{[*]} = \epsilon_{i+1} - \xi$ and $e_i = \epsilon_i - \xi$ and $\breve{c}_j = \frac{\Gamma(\delta+1)}{\Gamma(\gamma\delta+1)} \frac{\left[{}_{c}^{\smile}\partial_{\xi}^{\gamma\delta}\right] f(\xi)}{\left[{}_{c}^{\smile}\partial_{\xi}^{\delta}\right] f(\xi)}, \gamma = 2, 3, \ldots$.

5.1 Example 4.1: Bratu-type fractional boundary problem-engineering application

The Bratu-type fractional boundary problem is a boundary value problem with fractional differential equations. This problem extends the classical Bratu problem to the boundary value of a second-order ordinary differential equation. It is particularly relevant in engineering fields such as heat transport, chemical reaction kinetics, and combustion processes [47, 48].

$$\begin{cases} \left[{}_{\bar{c}}\mathcal{D}^{\delta}_{\delta_1}\right] u(\epsilon) + e^{u(\varepsilon)} = 0; 0 \le \epsilon \le 1, 1 \le \delta \le 2 \\ u(0) = 0; u(1) = 0, \end{cases} \tag{73}$$

In this context, the order δ can be any real number, not just an integer. The use of fractional derivatives allows for the simulation of phenomena such as memory effects and long-range interactions.

To solve such problems, specialized fractional calculus techniques, such as fractional integration and differentiation, as well as numerical methods tailored to fractional differential equations, are often necessary. Using the method defined in Ref. [47], we approximate the following polynomial up to two terms as:

$$u(\epsilon) = \sum_{i=0}^{2} u_i(\epsilon) = u_0(\epsilon) + u_1(\epsilon) + u_2(\epsilon). \tag{74}$$

where $u_0(\epsilon) = \sigma^{[2]}; u_1(\epsilon) = -\sigma^{[2]} + \frac{e^{\sigma^{[2]}}}{\Gamma(\delta+1)}\epsilon - \frac{e^{\sigma^{[2]}}}{\Gamma(\delta+1)}\epsilon^{\delta}$;

$u_1(\epsilon) = \frac{4^{-\delta} e^{\sigma^{[2]}} \left(e^{\sigma^{[2]}} \Gamma\left(\delta+\frac{1}{2}\right)\left(e^{\sigma^{[2]}} - \sigma^{[2]}\Gamma(\delta+2)\right)\right) - \sqrt{\pi}\sigma^{[2]}\Gamma(\delta+2)}{\delta\Gamma(\delta)\sqrt{\pi}\Gamma\left(\delta+\frac{1}{2}\right)\Gamma(\delta+2)}\epsilon + \frac{\sigma^{[2]} e^{\sigma^{[2]}}}{\Gamma(\delta+1)}\epsilon^{\delta} - \frac{e^{2\sigma^{[2]}}}{\Gamma(\delta+1)\Gamma(\delta+2)}\epsilon^{\delta+1} +$
$\frac{e^{2\sigma^{[2]}}}{\Gamma(2\delta+1)}\epsilon^{2\delta}$. To analyze the behavior of the Bratu-type boundary value problem, we solve:

$$\begin{aligned} u(\epsilon) = {} & \frac{e^{\sigma^{[2]}}}{\Gamma(\delta+1)}\epsilon - \frac{e^{\sigma^{[2]}}}{\Gamma(\delta+1)}\epsilon^{\delta} + \frac{4^{-\delta} e^{\sigma^{[2]}} \left(e^{\sigma^{[2]}} \Gamma\left(\delta+\frac{1}{2}\right)\left(e^{\sigma^{[2]}} - \sigma^{[2]}\Gamma(\delta+2)\right)\right) - \sqrt{\pi}\sigma^{[2]}\Gamma(\delta+2)}{\delta\Gamma(\delta)\sqrt{\pi}\Gamma\left(\delta+\frac{1}{2}\right)\Gamma(\delta+2)}\epsilon \\ & + \frac{\sigma^{[2]} e^{\sigma^{[2]}}}{\Gamma(\delta+1)}\epsilon^{\delta} - \frac{e^{2\sigma^{[2]}}}{\Gamma(\delta+1)\Gamma(\delta+2)}\epsilon^{\delta+1} + \frac{e^{2\sigma^{[2]}}}{\Gamma(2\delta+1)}\epsilon^{2\delta} \end{aligned} \tag{75}$$

Using $\epsilon_0 = 3.03$, we approximate our desired real root to be 0.0006. The analysis of the numerical schemes for solving this equation is given in **Tables 1** and **2**.

The numerical results are presented in **Table 1** using two different stopping criteria, $|f(\epsilon_i)|$ and $|\epsilon_{i+1} - \epsilon_i|$. **Table 1** includes the average CPU time (Ava-CPU), average number of iterations (Ava-it), and residual errors for various numerical schemes using fractional parameter values δ. An in-depth assessment of these indicators determines the numerical schemes that combine the best computational efficiency and solution accuracy, leading to the selection of the optimal techniques. The results indicate that the accuracy of the numerical schemes increases and the number of iterations decreases as the fractional parameter values increase from 0.1 to 1.0. **Table 1** also demonstrates that $MM^{[-]}$ and $MM^{[+]}$ outperforms existing fractional methods $MM^{[*]}$ and $MM^{[**]}$ in terms of Ave-it, Ave-CPU time, and residual error for solving the fractional Bratu-type problem.

The fractal behavior of iterative approaches for solving nonlinear equations is a fascinating phenomenon that reveals the complexities of numerical computations. Iterative algorithms utilize repeated steps to approximate the roots of nonlinear equations, often exhibiting fractal-like self-similar patterns. Through iterative refinement

Stopping criterion: $|f(\epsilon_i)|$							
Scheme	$\boldsymbol{\delta = 0.1}$	$\boldsymbol{\delta = 0.3}$	$\boldsymbol{\delta = 0.5}$	$\boldsymbol{\delta = 0.7}$	$\boldsymbol{\delta = 0.9}$	**Ave-It**	**Ave-CPU**
$MM^{[+]}$	3.43	5.15×10^{-1}	8.10×10^{-3}	1.76×10^{-4}	1.33×10^{-21}	18	6.7087
$MM^{[-]}$	0.03	1.19×10^{-2}	4.19×10^{-5}	1.08×10^{-7}	5.68×10^{-29}	17	5.0076
$MM^{[*]}$	0.75	9.17×10^{-1}	6.00×10^{-2}	3.02×10^{-4}	5.57×10^{-15}	35	3.2687
$MM^{[**]}$	0.77	2.77×10^{-1}	2.92×10^{-2}	1.05×10^{-3}	0.68×10^{-25}	41	3.2070
Stopping criterion: $|\epsilon_{i+1} - \epsilon_i|$							
Scheme	$\delta = 0.1$	$\delta = 0.3$	$\delta = 0.5$	$\delta = 0.7$	$\delta = 0.9$	Ave-It	Ave-CPU
$MM^{[+]}$	3.43	4.5×10^{-2}	7.0×10^{-3}	1.11×10^{-3}	3.0×10^{-22}	19	6.7007
$MM^{[-]}$	0.03	4.0×10^{-3}	4.09×10^{-4}	8.0×10^{-6}	6.5×10^{-26}	18	5.8066
$MM^{[*]}$	0.75	0.05	5.0×10^{-3}	1.03×10^{-2}	5.0×10^{-16}	36	3.0006
$MM^{[**]}$	0.77	0.07	2.01×10^{-1}	1.5×10^{-5}	6.5×10^{-19}	42	3.2995

Table 1.
Numerical results for Eq. (75) using $MM^{[+]}$–$MM^{[-]}$ and $MM^{[*]}$–$MM^{[**]}$ under two termination criteria.

Method	**Figure**	**Elapsed time**	**Divergent points**	**Convergent points**	**% Convergence**
$MM^{[+]}$	**Figure 7(a)**	153.876	1567.315	358432.685	99.5645
$MM^{[+]}$	**Figure 7(b)**	154.323	2891.421	357108.579	99.1934
$MM^{[+]}$	**Figure 7(c)**	132.456	3565.761	356434.239	98.1255
$MM^{[+]}$	**Figure 7(d)**	103.876	3982.913	356017.087	98.4398
$MM^{[-]}$	**Figure 7(e)**	55.2751	103.5490	359896.451	99.1256
$MM^{[-]}$	**Figure 7(f)**	65.8761	765.3120	359234.688	99.9264
$MM^{[-]}$	**Figure 7(g)**	76.2763	495.2130	359504.787	99.9167
$MM^{[-]}$	**Figure 7(h)**	85.5344	3475.678	356524.313	98.0347
$MM^{[*]}$	**Figure 7(i)**	170.247	554314.430	304568.570	15.1234
$MM^{[**]}$	**Figure 7(j)**	185.348	527385.120	307261.880	16.5454

Table 2.
Fractal analysis of $MM^{[+]}$–$MM^{[-]}$ and $MM^{[*]}$–$MM^{[**]}$ for solving Eq. (75).

of estimates, these approaches traverse zones of convergence and divergence, revealing complex structures that exhibit self-similarity at various scales. The Mandelbrot set, a classic example of fractals in mathematics, graphically illustrates this concept, highlighting the intricate boundaries between convergence and divergence for a specific iterative function. For more details on the dynamical behavior of iterative methods, refer to Ref. [49].

A grid of 600×600, centered at the origin, is used to create basins of attraction, with a total of 360,000 points employed to generate the dynamical planes within the square $[-2.0 \times 2.0]^2 \in \mathbb{C}$. Each root of $f(\epsilon) = 0$ is assigned a color, where the associated orbit of the iterative methods begins and converges to a fixed point for varying values of δ. The color map is set to Jet. As a stopping criterion, we select $|f(\epsilon_i)| < 0.0001$, with a maximum of 20 iterations.

Table 2 clearly shows that in terms of elapsed time (Elas-time), divergence points (Div points), convergence points (Con points), and percentage-convergence (Per-Con), the family of fractional numerical schemes $MM^{[-]}$ performs better than $MM^{[+]}$ and $MM^{[*]}$–$MM^{[**]}$. In the fractal of attractions shown in **Figure 8(a–j)**, bright colors indicate a lower number of iterations, while dark blue areas indicate divergence. **Figure 8(e–h)** demonstrates that, in terms of brightness and divergence zones, $MM^{[-]}$ outperforms $MM^{[+]}$ and $MM^{[*]}$–$MM^{[**]}$ for different parameter $\vartheta^{[*]}$ values, based on the information obtained in Section 3.

5.2 Example 4.2: Civil engineering application

The hanging object model, which includes systems such as the vertical spring-mass system, is widely used in engineering to understand and evaluate dynamic behavior. This model provides essential insights into oscillatory motion, which is critical for ensuring stability, preventing mechanical or structural failures, and supporting safe and reliable operation in various applications. Its relevance spans multiple engineering disciplines. In mechanical engineering, it is fundamental for vibration analysis and

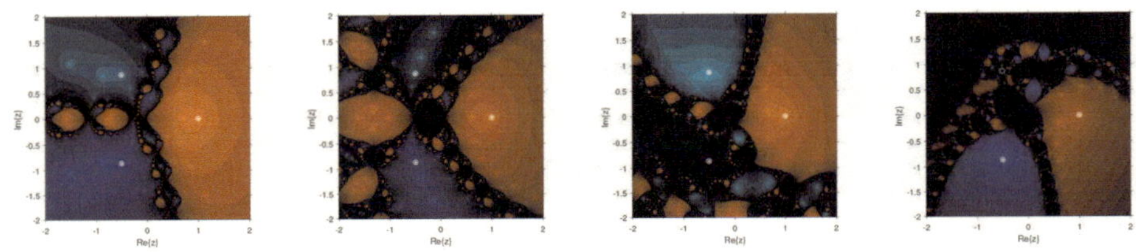

(a) Fractal graph of $MM^{[+]}$ generated for (75) with $\vartheta^{[*]} = 0$. (b) Fractal graph of $MM^{[+]}$ generated for (75) with $\vartheta^{[*]} = -1$. (c) Fractal graph of $MM^{[+]}$ generated for (75) with $\vartheta^{[*]} = 1.1i$. (d) Fractal graph of $MM^{[+]}$ generated for (75) with $\vartheta^{[*]} == 2i$.

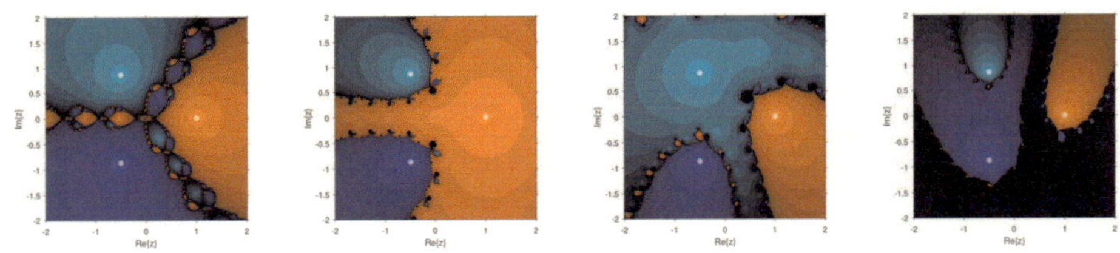

(e) Fractal graph of $MM^{[-]}$ generated for (75) with $\vartheta^{[*]} = 0$. (f) Fractal graph of $MM^{[-]}$ generated for (75) with $\vartheta^{[*]} = -1$ (g) Fractal graph of $MM^{[-]}$ generated for (75) with $\vartheta^{[*]} = 1.1i$ (h) Fractal graph of $MM^{[-]}$ generated for (75) with $\vartheta^{[*]} = -2i$

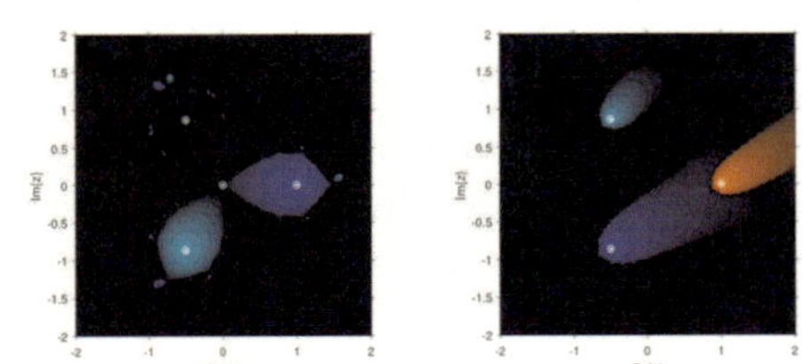

(i) Fractal graph of $MM^{[*]}$ generated for (75) with $\vartheta^{[*]} = 0.1$ (j) Fractal graph of $MM^{[**]}$ generated for (75) with $\vartheta^{[*]} = 0.1i$

Figure 8.
(a–j), shows the fractal analysis of the families of fractional schemes $MM^{[+]}$-$MM^{[-]}$, $MM^{[*]}$-$MM^{[**]}$ *generated for (75) with different values of* $\vartheta^{[*]}$.

suspension system design. Civil engineers apply it in the construction of stable structures, such as cable-stayed bridges, and in seismic analysis. In aerospace engineering, it contributes to the study of spacecraft component dynamics and flight stability. Automation experts employ it to develop control algorithms and balancing systems, while electrical engineers use it for MEMS devices and resonant circuits. Marine engineers apply these concepts to buoy dynamics and ship stabilization. The diverse applications of the hanging object model underscore its fundamental importance in creating robust, effective, and dependable engineering solutions.

To model the forces acting on a hanging object, we must compute the differential equations as follows:

$$\begin{cases} \dfrac{d\varkappa}{d\epsilon} - \left(\dfrac{d\varkappa}{d\epsilon}\right)^2 + 13\epsilon + 1 = 0, \\ \varkappa(0) = 1, \varkappa'(0) = \dfrac{e^2 - 1}{e^2 + 1}. \end{cases} \tag{76}$$

Utilizing the strategy suggested in Ref. [50], the polynomial is used to approximate the solution to this differential equation:

$$\varkappa(\epsilon) = -0.02590111180\epsilon^4 - 0.1066166681\epsilon^3 - 0.2099871708\epsilon^2 + 0.7615941560\epsilon \tag{77}$$

Using $\epsilon_0 = 0.3$, we have approximated our desired real root as 1.6609. The analysis of the numerical schemes for solving this differential equation is presented in **Tables 3** and **4**.

The numerical results are presented in **Table 3** using two different stopping criteria: $|\varkappa(\epsilon_i)|$ and $|\epsilon_{i+1} - \epsilon_i|$. The numerical schemes are evaluated based on computational efficiency and solution accuracy to identify the most effective techniques. The assessment shows that the accuracy of the numerical schemes improves and the number of iterations decreases as the fractional parameter values increase from 0.1 to 1.0. **Table 3** also demonstrates that $MM^{[-]}$ and $MM^{[+]}$ outperform the existing fractional methods $MM^{[*]}$ and $MM^{[**]}$ in terms of the average number of iterations (Ave-it), average CPU time (Ave-CPU), and residual error when solving the fractional Bratu-type problem. The fractal behavior of the iterative techniques $MM^{[-]}$ and $MM^{[+]}$ for solving (77) is analyzed in **Table 4** and visualized in **Figure 9(a–j)**.

Table 4 shows that, in terms of elapsed time (Elas-time), divergence points (Div points), convergence points (Con points), and percentage convergence (Per-Con), the fractional numerical scheme $MM^{[-]}$ outperforms $MM^{[+]}$ and $MM^{[*]}$–$MM^{[**]}$. The bright colors in the fractal of attractions in **Figure 9(a–j)** indicate a lower number of iterations, while the dark blue areas correspond to divergence. **Figure 9(e–h)** further demonstrate that, with respect to brightness and divergence regions, $MM^{[-]}$ outperforms $MM^{[+]}$ and $MM^{[*]}$–$MM^{[**]}$ across different $\vartheta^{[*]}$ parameter values, based on the results presented in Section 3.

5.3 Discussion of results and methods

This section provides an in-depth assessment of the performance of the proposed method, combining a dynamical systems perspective with quantitative numerical benchmarks.

Stopping criterion: $\lvert x(\epsilon_i)\rvert$							
Scheme	$\delta = 0.1$	$\delta = 0.3$	$\delta = 0.5$	$\delta = 0.7$	$\delta = 0.9$	**Ave-It**	**Ave-CPU**
$MM^{[+]}$	1.43	6.85×10^{-1}	5.16×10^{-3}	1.76×10^{-4}	9.19×10^{-21}	19	7.7543
$MM^{[-]}$	0.17	5.48×10^{-2}	2.12×10^{-5}	1.08×10^{-7}	5.08×10^{-29}	18	4.1246
$MM^{[*]}$	2.21	4.04×10^{-2}	8.01×10^{-3}	3.02×10^{-4}	9.88×10^{-15}	37	3.0987
$MM^{[**]}$	0.77	2.95×10^{-1}	1.91×10^{-2}	1.05×10^{-3}	7.67×10^{-25}	42	3.5432
Stopping criterion: $\lvert\epsilon_{i+1} - \epsilon_i\rvert$							
Scheme	$\delta = 0.1$	$\delta = 0.3$	$\delta = 0.5$	$\delta = 0.7$	$\delta = 0.9$	Ave-It	Ave-CPU
$MM^{[+]}$	7.43	9.5×10^{-2}	2.0×10^{-3}	9.79×10^{-3}	5.09×10^{-21}	20	6.8765
$MM^{[-]}$	0.03	5.0×10^{-3}	3.01×10^{-4}	7.06×10^{-5}	4.45×10^{-25}	18	5.8667
$MM^{[*]}$	0.13	0.03	1.0×10^{-4}	4.05×10^{-4}	7.22×10^{-15}	35	3.3406
$MM^{[**]}$	0.68	0.09	7.09×10^{-1}	1.5×10^{-5}	8.62×10^{-18}	47	4.2005

Table 3.
Numerical results for Eq. (77) using $MM^{[+]}$–$MM^{[-]}$ and $MM^{[*]}$–$MM^{[**]}$ under two termination criteria.

Method	Figure	Elapsed time	Divergent points	Convergent points	% Convergence
$MM^{[+]}$	**Figure 8(a)**	153.876	1567.315	358432.685	99.5646
$MM^{[+]}$	**Figure 8(b)**	154.323	2791.421	357208.579	99.2246
$MM^{[+]}$	**Figure 8(c)**	132.456	3565.761	356434.239	99.0095
$MM^{[+]}$	**Figure 8(d)**	103.876	3982.913	356017.913	98.8936
$MM^{[-]}$	**Figure 8(e)**	55.2751	103.5490	359896.451	99.9712
$MM^{[-]}$	**Figure 8(f)**	65.8761	765.3120	359234.688	99.7874
$MM^{[-]}$	**Figure 8(g)**	76.2763	495.2130	359504.213	99.8624
$MM^{[-]}$	**Figure 8(h)**	85.5344	3475.678	356524.322	99.5433
$MM^{[*]}$	**Figure 8(i)**	170.247	55431.430	304568.570	15.3976
$MM^{[**]}$	**Figure 8(j)**	185.348	52738.120	307261.880	85.3505

Table 4.
Fractal analysis of $MM^{[+]}$–$MM^{[-]}$ and $MM^{[*]}$–$MM^{[**]}$ for solving Eq. (77).

5.3.1 Dynamical behavior

Graphically, **Figures 2–7** illustrate the effectiveness of the fractional schemes $MM^{[+]}$ and $MM^{[-]}$ using parametric and dynamical planes. **Figures 2** and **3** depict convergence (red) and divergence (black) regions based on the control parameters, enabling optimal parameter selection. The dynamical planes in **Figures 4** and **5** show the structural dynamics of the convergence basins: well-connected and stable for parameters within the red regions and significantly sensitive or chaotic in the black

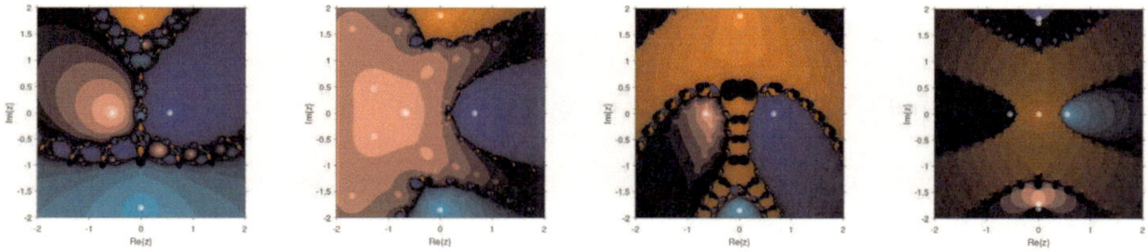

(a) Fractal graph of $MM^{[+]}$ generated for (76) with $\vartheta^{[*]} = 0$.
(b) Fractal graph of $MM^{[+]}$ generated for (76) with $\vartheta^{[*]} = -1$.
(c) Fractal graph of $MM^{[+]}$ generated for (76) with $\vartheta^{[*]} = 1.1i$.
(d) Fractal graph of $MM^{[+]}$ generated for (76) with $\vartheta^{[*]} == 2i$.

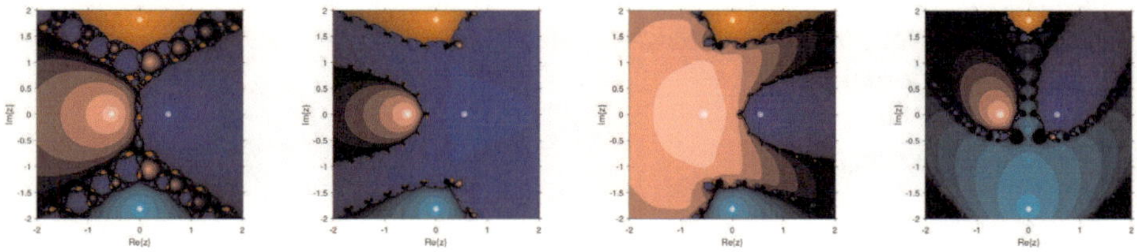

(e) Fractal graph of $MM^{[-]}$ generated for (76) with $\vartheta^{[*]} = 0$.
(f) Fractal graph of $MM^{[-]}$ generated for (76) with $\vartheta^{[*]} = -1$
(g) Fractal graph of $MM^{[-]}$ generated for (76) with $\vartheta^{[*]} = 1.1i$
(h) Fractal graph of $MM^{[-]}$ generated for (76) with $\vartheta^{[*]} = -2i$

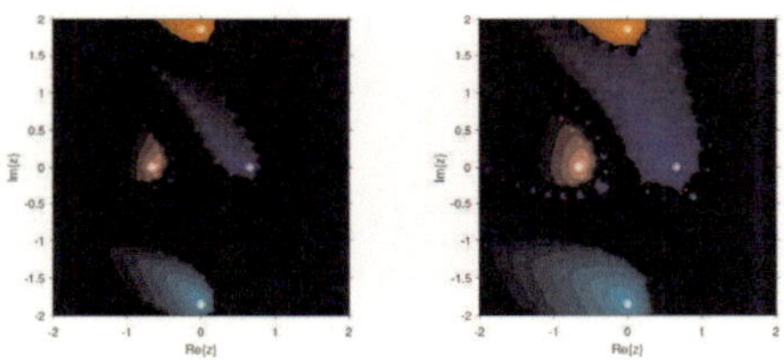

(i) Fractal graph of $MM^{[*]}$ generated for (76) with $\vartheta^{[*]} = 0.1$
(j) Fractal graph of $MM^{[**]}$ generated for (76) with $\vartheta^{[*]} = 0.1i$

Figure 9.
(a–j) shows the fractal analysis of the families of fractional schemes $MM^{[+]}$-$MM^{[-]}$, $MM^{[*]}$-$MM^{[**]}$ generated for (76) with different values of $\vartheta^{[*]}$.

regions. **Figures 6** and 7 display divergent iteration behavior, which directly confirms the instability resulting from unsuitable parameter choices. This visual evidence supports the theoretical stability analysis of the schemes and assists in parameter adjustment to ensure robust convergence.

5.3.2 Numerical performance comparison

To demonstrate the practical effectiveness of the proposed schemes $MM^{[+]}$ and $MM^{[-]}$, two nonlinear benchmark problems were considered and solved using our fractional approaches alongside classical fractional techniques, namely $MM^{[*]}$ and $MM^{[**]}$. The results are summarized in **Tables 1–4** and evaluated using several performance metrics:

- The residual error at the final iteration.
- The average number of iterations required to satisfy the convergence criteria.

- CPU time and elapsed time.
- The number of converging versus diverging initial points.
- In fractal analysis (basins of attraction), the percentage convergence calculated over a grid of initial starting points.

In both test engineering problems, our fractional techniques $MM^{[+]}$ and $MM^{[-]}$ consistently yield the lowest residual errors and require fewer average iterations while also operating significantly faster. In addition, they achieve higher convergence rates and demonstrate strong resistance to divergence, particularly in high-dimensional and sensitive problem domains.

5.3.3 Methodological insights

The higher performance is attributed to the combination of fractional-order dynamics and the critical-point-based initialization method. These advancements enable the approach to explore the solution space more effectively while maintaining numerical stability without requiring extensive adjustments. In contrast to existing methods, which may experience stagnation or sensitivity to parameter choices, the proposed schemes exhibit improved convergence properties.

5.3.4 Overall implications

These findings demonstrate the dual advantages of the proposed strategy:

1. The visual analysis tools support parameter selection and stability assessment.
2. The numerical metrics confirm more precise and faster convergence.

This confirms the proposed method as a reliable and efficient option for solving nonlinear equations, particularly in situations where existing methods fail or require substantial adjustment.

6. Conclusion

The Caputo-type fractional family of iterative algorithms, with convergence orders of $\delta + 1$, has been developed to solve nonlinear equations. Our numerical results confirm the theoretical convergence order, as established through a thorough convergence analysis. Using parametric and dynamical planes, we identified optimal parameter values for the fractional schemes $MM^{[+]}$ and $MM^{[-]}$. Engineering applications informed by the insights gained from the dynamical analysis are presented in **Tables 1–4**. Our investigation of fractal behavior (as detailed in **Tables 2** and **4**, and **Figures 2–9**) underscores the consistency of the newly developed methods. Notably, the fractional-order numerical schemes $MM^{[-]}$ consistently outperform $MM^{[+]}$ and $MM^{[*]}$–$MM^{[**]}$ across various fractional parameter values δ, in terms of the number of iterations, residual error, and computational convergence rates. **Tables 1** and **3** provide examples of numerical engineering applications, illustrating the efficacy and reliability of our iterative schemes compared to established techniques.

Key Outcomes and Advantages

- The use of fractional-order parameters enables more precise control over convergence behavior, expanding the design space of iterative techniques.
- The methods are computationally efficient while providing increased accuracy and robustness.
- Critical points in parametric and dynamical plane analysis serve as effective visual tools for identifying and optimizing the stability of the techniques.
- The proposed techniques outperform existing methods in terms of convergence time, precision, and resilience to divergence, particularly in high-dimensional and sensitive problem domains.

7. Limitations and challenges

Despite their strong performance, the schemes require appropriate selection of the fractional parameter to achieve optimal convergence. Although this can often be addressed through visual inspection or preliminary testing, the question of adaptive or fully automated parameter management remains largely unexplored. In addition, while the visualization techniques are informative, they may be computationally expensive for real-time processing or very large-scale systems.

8. Applications and relevance

The proposed approach has wide-ranging implications, with potential applications in computational physics, nonlinear dynamics, electrical circuit analysis, biological systems, and structural mechanics. Due to their ability to handle complex root structures with minimal a priori information, these methods are particularly well suited for inverse problems and parameter estimation tasks in scientific computing.

9. Future outlook

Looking ahead, several research directions could further deepen and expand the framework presented in this chapter. These include the following:

1. Developing adaptive and self-tuning algorithms for identifying suitable fractional parameters.
2. Integrating evolutionary or machine learning-based optimizers to improve robustness in noisy or unpredictable conditions.

3. Applying the approach to real-world inverse engineering problems, such as corrosion detection in power systems, where data uncertainty and nonlinearity are predominant.

4. Formulating advanced fractional iterative algorithms [51] that employ various fractional derivatives to achieve higher-order precision.

The fractional-order schemes presented in this work represent a novel and adaptable class of iterative approaches that balance theoretical rigor with practical performance. They bridge the gap between classical numerical analysis and emerging fractional dynamics, providing a solid foundation for future academic research and industrial applications.

Author details

Mudassir Shams[1,2,3] and Bruno Carpentieri[1*]

1 Faculty of Engineering, Free University of Bozen-Bolzano, Bolzano, Italy

2 Department of Mathematics, Faculty of Arts and Science, Balıkesir University, Balıkesir, Turkey

3 Department of Mathematics and Statistics, Riphah International University, Islamabad, Pakistan

*Address all correspondence to: bruno.carpientieri@unibz.it

References

[1] Miller KS, Ross B. An Introduction to the Fractional Calculus and Fractional Differential Equations. New York: John Wiley & Sons, Inc.; 1993

[2] Dzherbashyan MM, Nersesian AB. About application of some integro-differential operators. Doklady AkademiiNauk (Proceedings of the Russian Academy of Sciences). 1958; **121**(2):210-213

[3] Oldham KB, Spanier J. The Fractional Calculus: Theory and Application of Differentiation and Integration to Arbitrary Order. New York, NY, USA: Academic Press; 1974, Sons, New York, NY, USA, 1993

[4] Agarwal RP, Lakshmikantham V, Nieto JJ. On the concept of solution for fractional differential equations with uncertainty. Nonlinear Analysis: Theory Methods & Applications. 2010;**72**(6): 2859-2862

[5] Anderson DR, Ulness DJ. Newly defined conformable derivatives. Advances in Dynamical Systems and Applications. 2015;**10**(2):109-137. ISSN 0973-5321

[6] El-Wakil SA, Abulwafa EM. Formulation and solution of space-time fractional Boussinesq equation. Nonlinear Dynamics. 2015;**80**(1–2): 167-175

[7] Fazekas B, Goda TJ. Numerical stress solutions for the accurate calibration of hyper-viscoelastic material models of polymer foams. International Journal of Solids and Structures. 2020;**191**:390-400

[8] Yang XH, Zhuang YB, Zhu JX, Le JB, Cheng J. Recent progress on multiscale modeling of electrochemistry. Wiley Interdisciplinary Reviews: Computational Molecular Science. 2022; **12**(1):e1559

[9] Molina MG, Cabrera MA, Ezquer RG, Fernandez PM, Zuccheretti E. Digital signal processing and numerical analysis for radar in geophysical applications. Advances in Space Research. 2013; **51**(10):1870-1877

[10] Kashima K, Yamamoto Y. System theory for numerical analysis. Automatica. 2007;**43**(7):1156-1164

[11] Yang X-J, Tenreiro Machado JA, Hristov J. Nonlinear dynamics for local fractional burgers equation arising in fractal flow. Nonlinear Dynamics. 2016; **84**(1):3-7

[12] Ervin VJ, Roop JP. Variational solution of fractional advection dispersion equations on bounded domains in d, Numer. Methods Partial Differential Equation. 2007;**23**(2): 256-281

[13] Mathai AM, Haubold HJ. Fractional and multivariable calculus. In: Model Building and Optimization Problems; Springer Optimization and its Applications 122. Berlin, Germany: Springer; 2017

[14] Atanackovic TM, Pilipovic S, Stankovic B, Zorica D. Fractional Calculus with Applications in Mechanics: Wave Propagation, Impact and Variational Principles. London, UK: Wiley; 2014

[15] Saravi M, Saravi S. A new look at one of the Newton-like methods for solving non-linear equations. International Journal of Numerical Methods and Applications. 2011;**5**(1):31-43

[16] Shams M, Rafique N, Kausar N, Agrawal P, Park C, Mir NA. On iterative

technique for estimating all roots of non-linear equations and its system with applications in differential equations. Advances in Differences Equations. 2021;**2021**:480

[17] Lukić T, Ralević NM. Geometric mean Newton's method for simple and multiple roots. Applied Mathematics Letters. 2008;**21**(1):30-36

[18] Noor MA. New family of iterative methods for nonlinear equations. Applied Mathematics and Computation. 2007;**190**(1):553-558

[19] Özban AY. Some new variants of Newton's method. Applied Mathematics Letters. 2004;**17**(6):677-682

[20] Mir NA, Yasmin N, Rafiq N. Quadrature based two-step iterative methods for non-linear equations. General Mathematics. 2008;**16**(1):33-45

[21] Chun C. Iterative methods improving Newton's method by the decomposition method. Computers Mathematics with Applications. 2005;**50** (10–12):1559-1568

[22] Abbasbandy S. Improving Newton–Raphson method for nonlinear equations by modified Adomian decomposition method. Applied Mathematics and Computation. 2003;**145**(2–3):887-893

[23] Cordero A, Jordán C, Torregrosa JR. One-point Newton-type iterative methods: A unified point of view. Journal of Computational and Applied Mathematics. 2015;**275**:366-374

[24] Gander W. On Halley's iteration method. The American Mathematical Monthly. 1985;**92**(2):131-134

[25] Polyak B, Tremba A. New versions of Newton method: Step-size choice, convergence domain and under-determined equations. Optimization Methods and Software. 2020;**35**(6): 1272-1303

[26] Jain P. Steffensen type methods for solving non-linear equations. Applied Mathematics and Computation. 2007; **194**(2):527-533

[27] Dehghan M, Hajarian M. Some derivative free quadratic and cubic convergence iterative formulas for solving nonlinear equations. Computational Applied Mathematics. 2010;**29**:19-30

[28] Bisheh-Niasar M, Saadatmandi A. Some novel Newton-type methods for solving nonlinear equations. Boletim da Sociedade Paranaense de Matemática. 2020;**38**(3):111-123

[29] Farman M, Akgül A, Alshaikh N, Azeem M, Asad J. Fractional-order Newton–raphson method for nonlinear equation with convergence and stability analyses. Fractals. 2023;**31**(10):2340079

[30] Naveen K, Prakasha DG, Mofarreh F, Haseeb A. A study on solutions of fractional order equal-width equation using novel approach. Modern Physics Letters B. 2025;**39**(20):2550058

[31] Ali N, Waseem M, Safdar M, Akgül A, Tolasa FT. Iterative solutions for nonlinear equations via fractional derivatives: Adaptations and advances. Applied Mathematics in Science and Engineering. 2024;**32**(1):2333816

[32] Erfanifar R, Hajarian M, Sayevand K. A family of iterative methods to solve nonlinear problems with applications in fractional differential equations. Mathematical Methods in the Applied Sciences. 2024;**47**(4):2099-2119

[33] Torres-Hernandez AF, Brambila-Paz AF, Iturrarán-Viveros U, Caballero-Cruz

R. Fractional Newton–Raphson method accelerated with Aitken's method. Axioms. 2021;**10**(2):47

[34] Akgül A, Cordero A, Torregrosa JR. A fractional Newton method with 2αth-order of convergence and its stability. Applied Mathematics Letters. 2019;**98**: 344-351

[35] Cajori F. Historical note on the Newton-Raphson method of approximation. The American Mathematical Monthly. 1911;**18**(2):29-32

[36] Kumar P, Agrawal OP. An approximate method for numerical solution of fractional differential equations. Signal Processing. 2006; **86**(10):2602-2610

[37] Chang CW, Qureshi S, Argyros IK, Saraz KM, Hincal E. A modified fractional Newton's solver. Axioms. 2024;**13**(10):689

[38] Shams M, Carpentieri B. A new high-order fractional parallel iterative scheme for solving nonlinear equations. Symmetry. 2024;**16**(11):1452

[39] Chicharro FI, Cordero A, Garrido N, Torregrosa JR. Generating root-finder iterative methods of second order: Convergence and stability. Axioms. 2019;**8**(2):55

[40] Gdawiec K, Kotarski W, Lisowska A. Newton's method with fractional derivatives and various iteration processes via visual analysis. Numerical Algorithms. 2021;**86**(3):953-1010

[41] Odibat ZM, Shawagfeh NT. Generalized Taylor's formula. Applied Mathematics and Computation. 2007; **186**:286-293

[42] Fischer A, Izmailov AF, Jelitte M. Newton-type methods near critical solutions of piecewise smooth nonlinear equations. Computational Optimization and Applications. 2021;**80**(2):587-615

[43] Argyros IK, Hilout S. Computational Methods in Nonlinear Analysis. New Jersey, NJ, USA: World Scientific Publishing Company; 2013

[44] Liu L, Wang X. Eighth-order methods with high efficiency index for solving nonlinear equations. Applied Mathematics and Computation. 2010; **215**(9):3449-3454

[45] Candelario G, Cordero A, Torregrosa JR. Multipoint fractional iterative methods with $(2\alpha + 1)$th-order of convergence for solving nonlinear problems. Mathematics. 2020;**8**(3): 452

[46] Shams M, Kausar N, Agarwal P, Jain S, Salman MA, Shah MA. On family of the Caputo-type fractional numerical scheme for solving polynomial equations. Applied Mathematics in Science and Engineering. 2023;**31**(1): 2181959

[47] Sirisubtawee S, Kaewta S. New modified Adomian decomposition recursion schemes for solving certain types of nonlinear fractional two-point boundary value problems. International Journal of Mathematics and Mathematical Sciences. 2017;**2017**:1-20. Article ID: 5742965

[48] Chapra CS. Applied Numerical Methods with MATLAB for Engineers and Scientists. 6th ed. New York, NY, USA: McGraw-Hill; 2010

[49] Rafiq N, Akram S, Mir NA, Shams M. Study of dynamical behavior and stability of iterative methods for non-linear equations with application in engineering. In: Mathematical Problems in Engineering. United States: Wiley; 2020. 20 p. Article ID: 3524324

[50] Shams M, Kausar N, Yaqoob N, Arif N, Addis GM, et al. Techniques for finding analytical solution of generalized fuzzy differential equations with applications. Complexity. 2023;**2023**: 1-31

[51] Chu Y, Rafiq N, Shams M, Akram S, Mir NA, Kalsoom H. Computer methodologies for the comparison of some efficient derivative free simultaneous iterative methods for finding roots of nonlinear equations. CMC-Computers Materials & Continua. 2020;**66**(1):275-290

Chapter 5

Innovative Solution for the Legendre Equation: A Gateway to Efficient Laplace Equation Solving

Elyas Arsanjani Toroqi, Hashem Saberi Najafi and Seyed Reza Mirshafaei

Abstract

In this work, we propose a novel approach to address the challenging problem of solving the Legendre equation embedded within the Laplace equation for electric field analysis. Our method leverages the generalized differential transform method (GDTM), a numerical-analytical technique, to derive an approximate solution to the associated Sturm-Liouville problem. This intermediary step enables us to solve the Legendre equation with significantly enhanced precision. Validation of our solution involves comparison with the exact nth degree Legendre polynomial, demonstrating superior accuracy compared to traditional approximation methods when estimating the internal and external potential of a sphere across various spatial intervals. This study offers a robust tool for overcoming complexities in electric field calculations involving the Laplace and Legendre equations, thereby advancing the field of mathematical physics.

Keywords: generalized differential transform method, Legendre polynomial, Sturm-Lioville equation, Laplace equation, generalized Taylor's formula

1. Introduction

In the realm of mathematical research, we embark on an intriguing exploration of polynomials, a captivating subject that demands our attention. These complex algebraic expressions, with their intricate interplay of variables and coefficients, present a fascinating study. As we delve into this mathematical landscape, we uncover the hidden intricacies and the elegant relationships that define these mathematical entities. Polynomials, with their unique structure, offer a powerful tool to model and understand various phenomena. The connection between fundamental polynomial expressions and their application in solving intricate equations is evident, showcasing the seamless integration of theoretical foundations and practical problem-solving in the mathematical sciences. This chapter aims to unravel the mysteries within these expressions, exploring the fundamental principles that govern their behavior and the profound connections they forge within the realm of mathematics.

The exploration of fractional calculus and its applications has gained significant traction in the scientific community, particularly in the realms of physics and engineering.

Various methods have been employed to study fractional systems and equations, leading to notable advancements in these fields [1–4]. The field of numerical analysis has witnessed remarkable advancements in solving polynomial equations, as evidenced by recent groundbreaking studies. A notable work introduces a sophisticated fractional Caputo-type method, revolutionizing convergence rates without the need for supplementary polynomial evaluations [5]. This innovative approach emerges as a highly accurate and efficient alternative to conventional solutions. Another significant research endeavor proposes a refined iterative technique within the realm of fractional calculus, demonstrating its practical utility in civil and chemical engineering applications [6]. Graphical analysis unequivocally confirms its superior convergence speed relative to established fractional iterative methods. Furthermore, a third influential work unveils a family of fourth-order iterative techniques, meticulously designed for single and multiple root-finding [7]. This family consistently attains optimal convergence, surpassing previous studies, as corroborated by rigorous computational and numerical evaluations. Collectively, these contributions enrich the mathematical landscape, offering more efficient and specialized tools for the intricate task of polynomial root-finding.

One of the fundamental partial differential equations (PDEs) in physics and electronics is the Laplace equation, which holds immense importance in describing equilibrium scenarios [8–12]. This equation elegantly captures the state of balance, such as in heat transfer problems, where it determines the temperature distribution when thermal equilibrium is achieved. Moreover, the Laplace equation finds applications in chemical material science, enabling the analysis of density distribution in equilibrium conditions under the influence of electric and gravitational fields. The solution to this equation often involves the use of Legendre polynomials, as discussed in Refs. [8, 10, 11]. To illustrate its practical significance, consider the density of a chemical material, denoted as$u(x)$, and its corresponding output flux in a given region V:

$$\int_{\partial V} F\vec{n}ds = 0.$$

Here, F denotes the flux, which is inherently associated with the gradient of the function u:

$$F = -\alpha \nabla u, \quad \alpha > 0.$$

In accordance with the divergence theorem, we obtain the following result:

$$\int_{\partial V} F\vec{n}ds = \int_V div(F)dx = 0.$$

The region V under consideration can be of arbitrary shape and size, and when we extend the concept to three-dimensional Cartesian coordinates, the Laplace equation takes on a more general form. In this representation, the equation describes the equilibrium behavior in a three-dimensional space, providing a powerful tool for modeling and analyzing various physical phenomena:

$$\begin{aligned} div(F) &= 0, \\ div(\nabla u) &= 0, \\ \nabla^2 u = 0, \nabla^2 &= \frac{\partial^2}{\partial x^2} + \frac{\partial^2}{\partial y^2} + \frac{\partial^2}{\partial z^2}. \end{aligned}$$

For the second example, let E, ρ, and ϵ_0 represent the electric field, charge density, and permittivity of vacuum, respectively. Consequently, the mathematical model corresponding to these quantities is expressed as follows:

$$\nabla.E = \frac{\rho}{\epsilon_0}, \qquad (1)$$
$$\nabla \times E = 0.$$

When the curl of the field vanishes, it follows from Liouville's theorem that the field can be represented as the gradient of a scalar function u. This scalar function is referred to as the potential, whose gradient corresponds to the electrostatic field:

$$E = -\nabla u. \qquad (2)$$

Based on Eqs. (1) and (2), we deduce the following relationship:

$$\nabla^2 u = -\frac{\rho}{\epsilon_0}. \qquad (3)$$

Eq. (3) is known as the Poisson equation. In the absence of electric charge, that is, when $\rho = 0$, the Poisson equation reduces to the Laplace equation:

$$\nabla^2 u = 0. \qquad (4)$$

By employing spherical coordinates in Eq. (4), we derive the following expression:

$$\nabla^2 u = \frac{1}{r^2}\left[\frac{\partial}{\partial r}\left(r^2\frac{\partial u}{\partial r}\right) + \frac{1}{\sin\theta}\frac{\partial}{\partial\theta}\left(\sin\theta\frac{\partial u}{\partial\theta}\right) + \frac{1}{\sin^2\theta}\frac{\partial^2 u}{\partial\phi^2}\right] = 0. \qquad (5)$$

One approach to solving Eq. (5) involves the method of separation of variables. We assume that the potential function adheres to the following form:

$$u = R(r).P(\theta).Q(\phi). \qquad (6)$$

Here, r, θ, and ϕ signify the radial distance, the angle between a vector and the z-axis, and the angle of the vector's projection onto the xy-plane with the positive x-axis, respectively. Upon substituting Eq. (6) into Eq. (5) and applying the directional symmetry condition as a boundary condition, Eq. (5) is decomposed into three ordinary differential equations, each reflecting the symmetry of the solution in the respective direction:

$$\frac{d^2Q}{d\phi^2} = -m^2 Q,$$
$$r^2\frac{d^2R}{dr^2} + 2r\frac{dR}{dr} - n(n+1)R = 0, \qquad (7)$$
$$\frac{1}{\sin\theta}\frac{d}{d\theta}\left(\sin\theta\frac{dP}{d\theta}\right) + n(n+1)P = 0. \qquad (8)$$

Here, $Q(\phi)$ is a constant, and $m^2 = 0$. Moreover, m and n serve as parameters for solving differential equations in spherical coordinates. Introducing a new variable $x = \cos\theta$, Eq. (8) can be rewritten as follows:

$$\sin^2\theta\frac{d^2P}{dx^2} - 2\cos\theta\frac{dP}{dx} + n(n+1)P = 0. \tag{9}$$

The general solutions of Eqs. (7) and (9) are given by:

$$R(x) = c_1 r^n + c_2 r^{-(n+1)},$$
$$P(x) = c_1 P_1 + c_2 P_2,$$

where

$$P_1(x) = 1 + \sum_{q=1}^{\infty} (-1)^q \frac{n(n-2)\cdots(n-2q+2)(n+1)\cdots(n+2q-1)}{(2q)!} x^{2q},$$
$$P_2(x) = x + \sum_{q=1}^{\infty} (-1)^q \frac{(n-1)(n-3)\cdots(n-2q+1)(n+2)\cdots(n+2q)}{(2q+1)!} x^{2q+1}.$$

The solution to the Laplace equation can also be expressed using the Legendre polynomial of degree n, as elegantly described by Rodrigues' formula [10–13]. However, traditional methods of solving this equation are often intricate and time-consuming. In the upcoming Section 3, we introduce a powerful technique known as the generalized differential transform, which offers a more efficient and accurate approach to finding solutions.

One of the key advantages of this method lies in its versatility. It provides improved approximations for not only fractional differential equations but also ordinary and partial differential equations [14–18]. This transformative technique opens up new possibilities for solving a wide range of equations, making it an invaluable tool for mathematicians and scientists alike.

In the following sections, we will delve into the intricacies of this method, exploring its applications and the profound impact it has on the field of differential equations.

1.1 Background

In this section, we present a concise overview of the fundamental concepts and theorems that underpin our exploration of the Laplace equation and its solution using the generalized differential transform method (GDTM). By establishing these mathematical foundations, we aim to provide a comprehensive understanding of the techniques employed in our study.

Theorem 1.1: (*The uniqueness theorem*) Let V denote a volume enclosed by a surface S. If a charge density ρ is distributed throughout V and the potential u_S is specified on S, then the potential within the entirety of V is uniquely determined.

For a detailed discussion and rigorous proof, the reader is referred to [19].

Theorem 1.2: The solution of the Laplace equation is fundamentally reliant on the specifications of the boundary conditions.

For a more thorough exploration of the subject, readers are referred to the in-depth analyses presented in Refs. [20–22]. The Caputo fractional derivative of order α can be expressed as follows:

$$D^\alpha f(x) = \frac{1}{\Gamma(-\alpha+l)} \int_a^x (x-\tau)^{-\alpha+l-1} f^{(l)}(\tau) d\tau,$$

where $l-1<\alpha\le l, l\in Z^+$. For further insights and an in-depth exploration of the topic, readers are encouraged to consult the references provided in Refs. [1–3]. We introduce the concept of the generalized differential transform corresponding to the kth derivative of a function $f(x)$ in the following manner:

$$F_\alpha(k)=\frac{1}{\Gamma(\alpha k+1)}\left[(D^\alpha)^k f(x)\right]_{x=x_0}, \tag{10}$$

where $0<\alpha\le 1$ and $(D^\alpha)^k=D^\alpha.D^\alpha.\cdots.D^\alpha(k-times)$. Furthermore, the inverse differential transform of $F_\alpha(k)$ is given by:

$$f(x)=\sum_{k=0}^{\infty}F_\alpha(k)(x-x_0)^{\alpha k}. \tag{11}$$

By replacing Eq. (10) in Eq. (11) and utilizing the generalized Taylor's formula, as presented in Ref. [23], the following result is obtained:

$$f(x)=\sum_{k=0}^{\infty}F_\alpha(k)(x-x_0)^{\alpha k}=\sum_{k=0}^{\infty}\frac{(x-x_0)^{\alpha k}}{\Gamma(\alpha k+1)}\left((D^\alpha)^k f\right)(x_0).$$

By applying Theorem (4) from Ref. [23], we obtain:

$$f(x)\cong\sum_{k=0}^{t}F_\alpha(k)(x-x_0)^{\alpha k}. \tag{12}$$

When t becomes adequately large, the subsequent theorems assist in addressing fractional differential equations.

Theorem 1.3: Suppose $f(x)$ is expressed as the sum or difference of two functions, $g(x)$ and $h(x)$. Then, under the transformation defined by $F_\alpha(k)$, the corresponding relationship becomes $F_\alpha(k)=G_\alpha(k)\pm H_\alpha(k)$, valid for $0<\alpha\le 1$.

Theorem 1.4: If $f(x)=cg(x)$ for some constant $c\in\mathbb{R}$, then $F_\alpha(k)=cG_\alpha(k)$, where $0<\alpha\le 1$.

Theorem 1.5: If the function $f(x)$ is defined as the fractional derivative of order of another function $g(x)$, i.e., $f(x)=D^\alpha g(x)$, then the transformed representation of $f(x)$, denoted by $F_\alpha(k)$, can be expressed in terms of $G_\alpha(k)$, the transformed representation of $g(x)$, as follows:

$$F_\alpha(k)=\frac{\Gamma(\alpha(k+1)+1)}{\Gamma(\alpha k+1)}G_\alpha(k+1),$$

where Γ represents the Gamma function, and $0<\alpha\le 1$.

Theorem 1.6: Suppose $f(x)$ is given by the fractional derivative of order β of a function $g(x)$, i.e., $f(x)=D^\beta g(x)$, where $l-1<\beta\le l$ for some integer l. Furthermore, assume that the function $g(x)$ satisfies the prerequisites outlined in Theorems (2–5) of the reference [24]. Under these conditions, the transformed representation of $f(x)$, denoted as $F_\alpha(k)$, is given by: $F_\alpha(k)=\frac{\Gamma(\alpha k+\beta+1)}{\Gamma(\alpha k+1)}G_\alpha\left(k+\frac{\beta}{\alpha}\right)$ where $0<\alpha\le 1$.

Detailed proofs of the results can be accessed in the reference [24].

2. Numerical results and discussion

In this section, we present the numerical solutions to Eq. (9) using Rodrigues' formula and other established methods. Additionally, we employ the Sturm-Liouville equation and the generalized differential transform method (GDTM) to obtain more accurate approximations.

Example 3.1. Consider a spherical capacitor, as illustrated in **Figure 1**, comprising two metallic hemispheres, each with a radius of 1 foot, separated by a narrow slit for isolation purposes. The upper hemisphere is maintained at a voltage of 110 V, while the lower hemisphere is grounded. The boundary conditions for this setup are as follows:

$$f(\theta) = \begin{cases} 110, & 0 \le \theta < \frac{\pi}{2}, \\ 0, & \frac{\pi}{2} < \theta \le \pi. \end{cases}$$

The inner and outer potentials of the sphere are expressed as follows, respectively:

$$u_n(r,\theta) = A_n r^n P_n(\cos\theta), \tag{13}$$

$$u_n(r,\theta) = \frac{B_n}{r^{n+1}} P_n(\cos\theta), \tag{14}$$

for $n = 0,1,2, \ldots$. The Legendre polynomials are denoted by $P_n(\cos\phi)$. A series expansion as described in Eq. (13) is considered as follows:

$$u(r,\theta) = \sum_{n=0}^{\infty} A_n r^n P_n(\cos\theta), \quad r \le R. \tag{15}$$

Given that the sphere S is defined by $r = R$, the Dirichlet condition is satisfied for Eq. (15) (see Eq. (9) in Section 12.11 of [11]). Consequently, we have:

$$u(R,\theta) = \sum_{n=0}^{\infty} A_n R^n P_n(\cos\theta) = f(\theta), \tag{16}$$

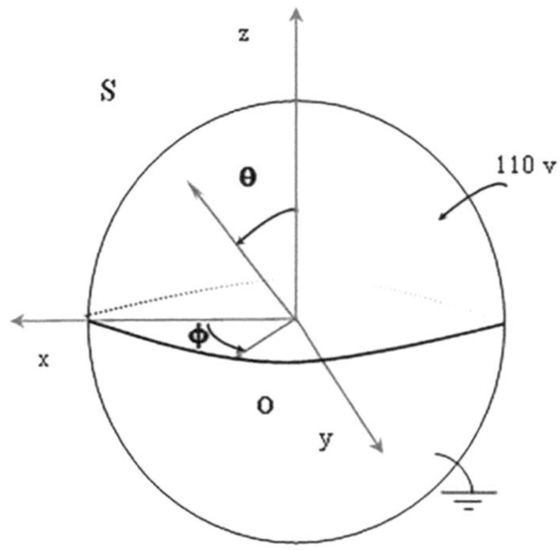

Figure 1.
The spherical capacitor.

where Eq. (16) corresponds to the Fourier-Legendre series representation of $f(\theta)$. Furthermore, from Eq. (7) in Section 11.9 of [11], it follows that:

$$A_n R^n = \frac{2n+1}{2}\int_{-1}^{1} \tilde{f}(\omega) P_n(\omega) d\omega.$$

Here, $\tilde{f}(\theta)$ denotes $f(\theta)$. Assuming $\omega = \cos\theta$, the integration limits from -1 to 1 correspond to $\theta = \pi$ and $\theta = 0$, respectively. Thus, we can rewrite the expression as follows:

$$A_n = \frac{2n+1}{2R^n}\int_0^{\pi} f(\theta) P_n(\cos\theta)\sin\theta \; d\theta, \quad n = 0,1,2, \dots . \tag{17}$$

Furthermore, it follows from Eq. (14) that:

$$u(r,\theta) = \sum_{n=0}^{\infty} \frac{B_n}{r^{n+1}} P_n(\cos\theta), \quad r \geq R.$$

By referencing Eqs. (8), (9), and (10) in Section 12.11 of [11], the following result is obtained:

$$B_n = \left(\frac{2n+1}{2}\right) R^{n+1} \int_0^{\pi} f(\theta) P_n(\cos\theta)\sin\theta \; d\theta, \quad n = 0,1,2, \dots .$$

Since $R = 1$, Eq. (17) can be rewritten as follows:

$$A_n = \left(\frac{2n+1}{2}\right) 110 \int_0^{\frac{\pi}{2}} P_n(\cos\theta)\sin\theta \; d\theta = \left(\frac{2n+1}{2}\right) 110 \int_0^{1} P_n(\omega) d\omega.$$

As per Section 5.2 in Ref. [11], we derive:

$$A_n = 55(2n+1)\sum_{m=0}^{M} \frac{(2n-2m)!}{2^n m!(n-m)!(n-2m)!}\int_0^1 \omega^{n-2m} \; d\omega.$$

For even values of n, M is defined as $\frac{n}{2}$, while for odd values of n, $M = \frac{n-1}{2}$. This definition applies for $n = 0,1,2, \dots$. The following result is obtained:

$$A_0 = 55, \; A_1 = \frac{165}{2}, \; A_2 = 0, \; A_3 = -\frac{385}{8}, \dots . \tag{18}$$

Upon substitution of Eq. (18) into Eq. (15), the resulting expression can be derived as follows:

$$u(r,\theta) = 55 + \left(\frac{165}{2}\right) r P_1(\cos\theta) - \left(\frac{385}{8}\right) r^3 P_3(\cos\theta) + \dots .$$

It is worth noting that the even-indexed coefficients of A_n, for $n = 2,4,6, \dots$, vanish. With $R = 1$, it is deduced that A_n equals B_n. Consequently, the equality of the inner and outer potentials of the sphere is established, as expressed below:

$$u(r,\theta) = \frac{55}{r} + \left(\frac{165}{2r^2}\right)P_1(\cos\theta) - \left(\frac{385}{8r^4}\right)P_3(\cos\theta) + \left(\frac{605}{16r^6}\right)P_5(\cos\theta) - \left(\frac{4125}{128r^8}\right)P_7(\cos\theta) + \left(\frac{7315}{256r^{10}}\right)P_9(\cos\theta) + \dots \,. \tag{19}$$

The Legendre polynomials of degree n, represented by $P_0, P_1, P_3, \dots$, can be computed using Rodrigues' formula:

$$P_n(\omega) = \frac{1}{2^n n!}\frac{d^n}{d\omega^n}\left(\omega^2 - 1\right)^n.$$

Therefore, we have:

$$P_0(\omega) = 1, P_1(\omega) = \omega, P_3(\omega) = \frac{1}{2}\left(5\omega^3 - 3\omega\right), P_5(\omega) = \frac{63}{8}\omega^5 - \frac{35}{4}\omega^3 + \frac{15}{8}\omega,$$
$$P_7(\omega) = \frac{429}{16}\omega^7 - \frac{693}{16}\omega^5 + \frac{315}{16}\omega^3 - \frac{35}{16}\omega, \tag{20}$$
$$P_9(\omega) = \frac{12155}{128}\omega^9 - \frac{6435}{32}\omega^7 + \frac{9009}{64}\omega^5 - \frac{1155}{32}\omega^3 + \frac{315}{128}\omega, \dots .$$

By substituting Eq. (20) into Eq. (19) and setting $r = 1$ and $\omega = \cos(\theta)$, the potential of the sphere can be derived using Rodrigues' formula, as shown below:

$$u(1,\theta) = 55 + \left(\frac{165}{2}\right)\omega - \left(\frac{385}{8}\right)\left(\frac{1}{2}\left(5\omega^3 - 3\omega\right)\right) + \left(\frac{605}{16}\right)\left(\frac{63}{8}\omega^5 - \frac{35}{4}\omega^3 + \frac{15}{8}\omega\right) - \left(\frac{4125}{128}\right)\left(\frac{429}{16}\omega^7 - \frac{693}{16}\omega^5 + \frac{315}{16}\omega^3 - \frac{35}{16}\omega\right) + \left(\frac{7315}{256}\right)\left(\frac{12155}{128}\omega^9 - \frac{6435}{32}\omega^7 + \frac{9009}{64}\omega^5 - \frac{1155}{32}\omega^3 + \frac{315}{128}\omega\right).$$

Due to the presence of the generalized differential transform $\sin^2\theta$ in the denominator, GDTM fails to solve Eq. (9) because this transform evaluates to zero at certain steps. Therefore, we employ the Sturm-Liouville equation to address the problem:

$$\frac{d}{d\omega}\left[h(\omega)\frac{dP}{d\omega}\right] + [i(\omega) + \lambda j(\omega)]P = 0.$$

Let $i = 0, j = 1$. Assuming $\omega = \cos\theta$ and $h(\omega) = 1 - \omega^2$, Eq. (9) can be transformed using the Sturm-Liouville equation as follows:

$$h(\omega)\frac{d^2P}{d\omega^2} - 2\omega\frac{dP}{d\omega} + \lambda P = 0, \quad \lambda = n(n+1). \tag{21}$$

Using Rodrigues' formula, we define the following initial conditions for $n = 0,1,3, \dots, 9$:

$$P_0(1) = 1, P'_0(1) = 0, \tag{22}$$

$$P_1(1) = 1, P'_1(1) = 1, \tag{23}$$

$$P_3(1) = 1, P_3'(1) = 6, \tag{24}$$

$$P_5(1) = 1, P_5'(1) = 15, \tag{25}$$

$$P_7(1) = 1, P_7'(1) = 28, \tag{26}$$

$$P_9(1) = 1, P_9'(1) = 45. \tag{27}$$

Assuming $\alpha = 1$ and $\beta = 2$, and in accordance with the definition of the Caputo fractional derivative along with Eq. (10), the generalized differential transforms corresponding to Eqs. (21) and (22) are obtained as follows:

$$P_1^0(k+2) = 0,$$
$$P_1^0(0) = 1, P_1^0(1) = 0.$$

Consequently, the solution $P_0(\omega)$ is determined up to $O\left((\omega-1)^0\right)$.

$$P_0(\omega) = 1.$$

In this context, $O((\omega-1)^0)$ signifies the truncation error, whereas $(\omega-1)^0$ corresponds to the initial term of the solution series generated by GDTM. Through the application of Theorems 1.5 and 1.6 for transforming the first and second terms of Eq. (21), and utilizing Theorems 1.3 and 1.4 for performing addition/subtraction and multiplication by λ, the generalized differential transform of Eq. (21) and the associated initial conditions (23)–(27) are determined for $n = 1, 3, \ldots, 9$:

$$P_1^1(k+2) = 2\frac{\Gamma(k+1)}{h(x)\Gamma(k+3)}\left[\omega\frac{\Gamma(k+2)}{\Gamma(k+1)}P_1^1(k+1) - P_1^1(k)\right], \tag{28}$$
$$P_1^1(0) = 1, P_1^1(1) = 0,$$

$$P_1^3(k+2) = \frac{\Gamma(k+1)}{h(x)\Gamma(k+3)}\left[2\omega\frac{\Gamma(k+2)}{\Gamma(k+1)}P_1^3(k+1) - 12P_1^3(k)\right], \tag{29}$$
$$P_1^3(0) = 1, P_1^3(1) = 0,$$

$$P_1^5(k+2) = \frac{\Gamma(k+1)}{h(x)\Gamma(k+3)}\left[2\omega\frac{\Gamma(k+2)}{\Gamma(k+1)}P_1^5(k+1) - 30P_1^5(k)\right], \tag{30}$$
$$P_1^5(0) = 1, P_1^5(1) = 0,$$

$$P_1^7(k+2) = \frac{\Gamma(k+1)}{h(x)\Gamma(k+3)}\left[2\omega\frac{\Gamma(k+2)}{\Gamma(k+1)}P_1^7(k+1) - 56P_1^7(k)\right], \tag{31}$$
$$P_1^7(0) = 1, P_1^7(1) = 0,$$

and

$$P_1^9(k+2) = \frac{\Gamma(k+1)}{h(x)\Gamma(k+3)}\left[2\omega\frac{\Gamma(k+2)}{\Gamma(k+1)}P_1^9(k+1) - 90P_1^9(k)\right], \tag{32}$$
$$P_1^9(0) = 1, P_1^9(1) = 0.$$

For each n, by considering $k = 0,1,2, \ldots$ in Eqs. (28)–(32) simultaneously and subsequently substituting the previous coefficients for $F_\alpha(k)$ and $\omega_0 = 1$ (which follows from the initial condition $\theta = 0$) in place of x_0 within Eq. (12), the Legendre polynomials $P_1(\omega), P_3(\omega), \dots, P_9(\omega)$ are determined, as expressed below:

$$P_1(\omega) = 1 - \left(\frac{1}{h}\right)(\omega-1)^2 - \left(\frac{2\omega}{3h^2}\right)(\omega-1)^3 + \left(\frac{1}{6h^2} - \frac{\omega^2}{3h^3}\right)(\omega-1)^4, \tag{33}$$

$$P_3(\omega) = 1 - \left(\frac{6}{h}\right)(\omega-1)^2 - \left(\frac{4\omega}{h^2}\right)(\omega-1)^3 + \left(\frac{6}{h^2} - \frac{2\omega^2}{h^3}\right)(\omega-1)^4, \tag{34}$$

$$P_5(\omega) = 1 - \left(\frac{15}{h}\right)(\omega-1)^2 - \left(\frac{10\omega}{h^2}\right)(\omega-1)^3 + \left(\frac{75}{2h^2} - \frac{5\omega^2}{h^3}\right)(\omega-1)^4, \tag{35}$$

$$P_7(\omega) = 1 - \left(\frac{28}{h}\right)(\omega-1)^2 - \left(\frac{56\omega}{3h^2}\right)(\omega-1)^3 + \left(\frac{392}{3h^2} - \frac{28\omega^2}{3h^3}\right)(\omega-1)^4, \tag{36}$$

$$P_9(\omega) = 1 - \left(\frac{45}{h}\right)(\omega-1)^2 - \left(\frac{30\omega}{h^2}\right)(\omega-1)^3 + \left(\frac{675}{2h^2} - \frac{15\omega^2}{h^3}\right)(\omega-1)^4. \tag{37}$$

It should be emphasized that incorporating additional terms in the series solution may lead to increased errors. Therefore, we limit the solution $P_n(\omega)$ to $O\left((\omega-1)^4\right)$. Furthermore, setting $\theta = 0$ alters Eq. (21), rendering it unsuitable as a Sturm-Liouville equation. By substituting the aforementioned equations into Eq. (19), we obtain the solutions of the Laplace equation using GDTM for $\theta \in [0.1745329252, 1.570796327]$ as follows:

$$\begin{aligned} u(1,\theta) = 55 &+ \frac{165}{2}\left[1 - \left(\frac{1}{h}\right)(\omega-1)^2 - \left(\frac{2\omega}{3h^2}\right)(\omega-1)^3 + \left(\frac{1}{6h^2} - \frac{\omega^2}{3h^3}\right)(\omega-1)^4\right] \\ &- \frac{385}{8}\left[1 - \left(\frac{6}{h}\right)(\omega-1)^2 - \left(\frac{4\omega}{h^2}\right)(\omega-1)^3 + \left(\frac{6}{h^2} - \frac{2\omega^2}{h^3}\right)(\omega-1)^4\right] \\ &+ \frac{605}{16}\left[1 - \left(\frac{15}{h}\right)(\omega-1)^2 - \left(\frac{10\omega}{h^2}\right)(\omega-1)^3 + \left(\frac{75}{2h^2} - \frac{5\omega^2}{h^3}\right)(\omega-1)^4\right] \\ &- \frac{4125}{128}[1 - \left(\frac{28}{h}\right)(\omega-1)^2 - \left(\frac{56\omega}{3h^2}\right)(\omega-1)^3 + \left(\frac{392}{3h^2} - \frac{28\omega^2}{3h^3}\right)(\omega-1)^4] \\ &+ \frac{7315}{256}\left[1 - \left(\frac{45}{h}\right)(\omega-1)^2 - \left(\frac{30\omega}{h^2}\right)(\omega-1)^3 + \left(\frac{675}{2h^2} - \frac{15\omega^2}{h^3}\right)(\omega-1)^4\right], \end{aligned} \tag{38}$$

where θ is shown in Radian. **Table 1** presents a comprehensive comparison of the approximations for the Legendre polynomial of degree 9 using four different methods: the generalized differential transform method (GDTM), Rodrigues' formula, Runge–Kutta Fehlberg method of fourth-fifth order (RKF45), and Taylor's series. The RKF45 method is a sophisticated Runge–Kutta approach, as detailed in Refs. [25, 26]. **Figure 2** illustrates the approximations and associated errors for the mentioned methods within the interval $\theta \in [0.01745329252, 0.1745329252]$. It is worth noting that GDTM demonstrates certain limitations in the initial segment of the interval. However, **Figures 3** and **4** provide further insights, demonstrating that the GDTM-derived values closely align with the ideal potential of 110 V and consistently result in the lowest error. As expected, the potential decreases significantly as we approach isolation, which in turn leads to an increase in error for all methods. **Figure 4** reveals that

θ	GDTM	Rodrigues	RKF45	Taylor's series
0.1745329252	119.6395530	115.0677953	108.3734725	108.3734725
0.3490658504	112.4109645	108.7032745	82.2353712	82.2353714
0.5235987758	113.3224144	120.7591281	78.0034333	84.1951037
0.6981317008	110.2591195	122.3741211	62.2939333	68.4800347
0.8726646262	111.8103094	141.8542600	77.2680049	83.3120615
1.047197551	114.8379630	161.4808985	154.8091511	160.3441611
1.221730477	109.3999094	169.3100614	137.5806397	142.1331072
1.396263402	91.53486009	159.5602254	95.5078245	98.5605815

Table 1.
Comparison GDTM with other methods in $\theta \in [0.1745329252, 1.396263402]$.

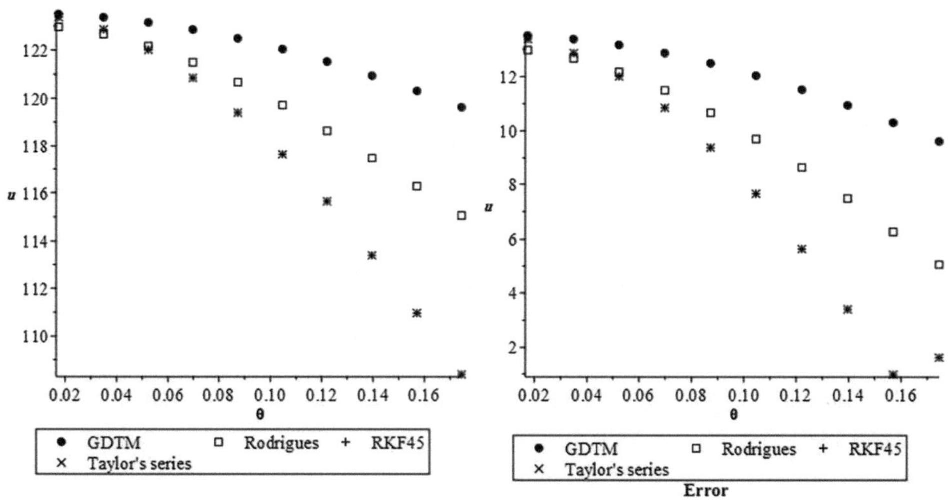

Figure 2.
Comparison GDTM with other methods (b) error in $\theta \in [0.01745329252, 0.1745329252]$.

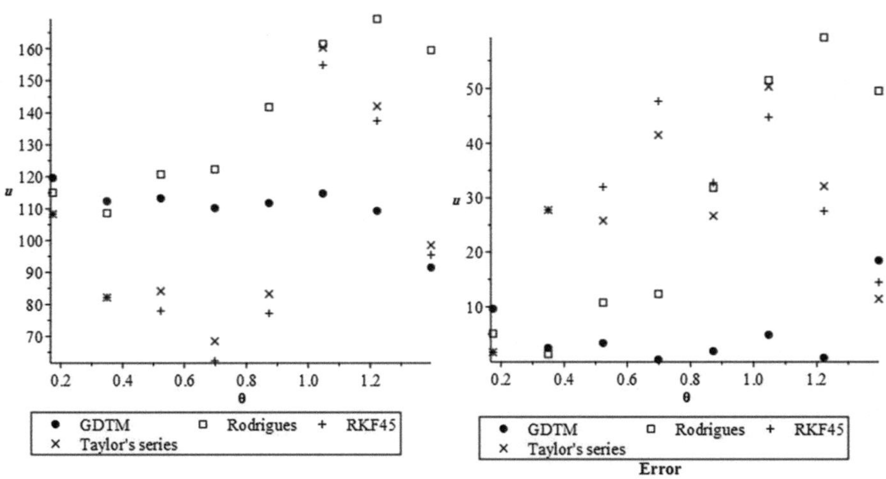

Figure 3.
Comparison GDTM with other methods (b) error in $\theta \in [0.1745329252, 1.396263402]$.

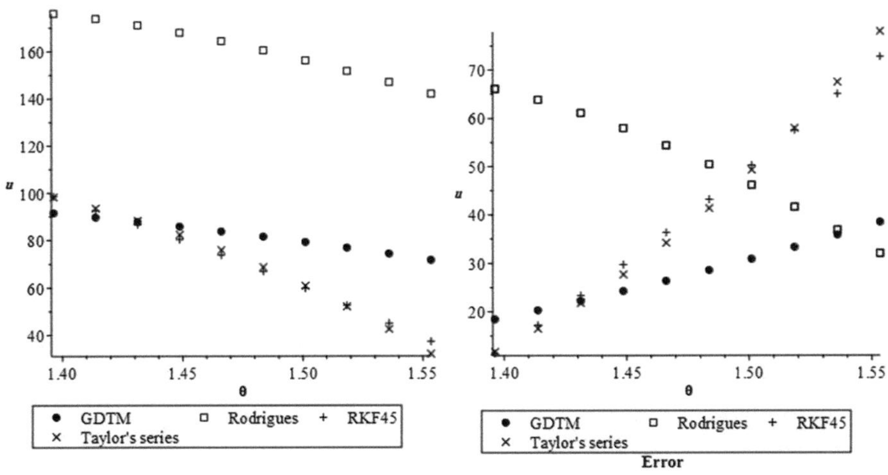

Figure 4.
Comparison GDTM with other methods (b) error in $\theta \in [1.396263402, 1.553343034]$.

the potential decline in GDTM is comparable to that of RKF45 and Taylor's series. These findings highlight the strengths and limitations of each method, providing valuable guidance for selecting the most suitable approach based on the specific problem at hand.

The GDTM significantly enhances computational efficiency in calculating series coefficients through its recursive approach. This method notably reduces the computational burden compared to traditional techniques that rely on solving extensive systems of equations, such as finite difference methods. The recursive nature of GDTM allows for the efficient computation of high-degree polynomials without the need for extensive matrix operations.

One of the key advantages of GDTM is its scalability with increasing polynomial degree. While methods like the Frobenius or finite difference methods can become computationally intensive for high-degree polynomials, GDTM maintains its efficiency due to its series-based recursive algorithm. This scalability ensures that the computational costs do not escalate significantly as the polynomial degree increases.

Furthermore, GDTM's ability to handle fractional derivatives enhances its applicability to complex problems while minimizing the need for additional computational resources. A key feature of the method is its recursive computation of coefficients, which facilitates parallelization and makes it highly suitable for high-performance computing environments. This parallelization is especially advantageous for large-scale problems with limited computational resources, enabling efficient use of available resources. In terms of memory usage, GDTM offers a more efficient approach than other methods. Rather than storing large matrices, GDTM focuses on storing a series of coefficients, thereby reducing memory requirements significantly. This memory efficiency is crucial for problems with high degrees of freedom, such as those involving high-degree Legendre polynomials, where memory constraints can pose significant challenges.

Moreover, GDTM demonstrates substantial potential for future research, offering multiple avenues for development and addressing current gaps. One promising direction involves extending GDTM to fractional differential equations. By leveraging GDTM's capacity to handle fractional derivatives, researchers could tackle intricate challenges in diverse fields such as electromagnetism, fluid dynamics, bioengineering,

DOI: http://dx.doi.org/10.5772/intechopen.1009778

and finance. For example, GDTM could model the behavior of complex materials under fluctuating electric fields. It could also provide solutions to heat conduction problems in irregularly shaped domains. These advancements highlight the versatility and applicability of GDTM in solving real-world issues.

Additionally, extending GDTM to nonlinear systems could unlock new solutions in physics, engineering, and biology. Building on its versatility, another promising direction is the integration of GDTM with machine learning, leveraging its recursive nature to develop hybrid methods for solving high-dimensional differential equations. In addition to expanding its application domains, developing adaptive GDTM methods that adjust the number of terms in the series expansion based on error estimates could improve both efficiency and accuracy, particularly for problems with varying complexity. These future directions highlight GDTM's potential to advance both theoretical and applied research in differential equations.

3. Conclusions

In this study, we successfully applied the generalized differential transform method (GDTM) to derive Lagrange polynomials, introducing a new approach for solving the Laplace equation. Although the inclusion of the variable h increases computational complexity, the GDTM approximation exhibits significantly higher accuracy than conventional techniques such as the Runge–Kutta Fehlberg method of fourth-fifth order or Taylor's series. Our error analysis confirms that results derived using alternative approaches often diverge considerably from the ideal potential, whereas GDTM consistently produces estimates that closely approximate the desired values over a wide range of intervals.

These findings advance the development of robust and precise methods for solving differential equations, particularly in physics and engineering contexts. Future investigations into enhancing GDTM's capabilities—such as combining it with complementary numerical techniques—could yield even more powerful tools for addressing intricate mathematical challenges.

Author details

Elyas Arsanjani Toroqi[1*†], Hashem Saberi Najafi[1] and Seyed Reza Mirshafaei[2]

1 Department of Applied Mathematics, Faculty of Mathematics Lahijan Branch Islamic Azad University, Lahijan, Iran

2 Department of Mathematics and Management, Roudsar and Amlash Branch, Islamic Azad University, Roudsar, Iran

*Address all correspondence to: arsanjanie@yahoo.com

† The author, designated as the primary contributor, spearheaded the composition and development of the chapter.

References

[1] Kilbas AA, Srivastava HM, Trujillo JJ. Theory and Applications of Fractional Differential Equations. Amsterdam, The Netherland: Elsevier; 2006

[2] Oldham K, Spanier J. The Fractional Calculus Theory and Applications of Differentiation and Integration to Arbitrary Order. Mineola, New York: Elsevier; 1974

[3] Podlubny I. Fractional Differential Equations. San Diego, United States of America: Academic Press; 1999

[4] Najafi HS, Sheikhani AR, Ansari A. Stability analysis of distributed order fractional differential equations. Inabstract and Applied Analysis. 2011; **2011**(1):175323). Hindawi Publishing Corporation. DOI: 10.1155/2011/175323

[5] Shams M, Kausar N, Samaniego C, Agarwal P, Ahmed SF, Momani S. On efficient fractional Caputo-type simultaneous scheme for finding all roots of polynomial equations with biomedical engineering applications. Fractals. 2023; **31**(04):2340075. DOI: 10.1142/S0218348X23400753

[6] Shams M, Kausar N, Agarwal P, Jain S, Salman MA, Shah MA. On family of the Caputo-type fractional numerical scheme for solving polynomial equations. Applied Mathematics in Science and Engineering. 2023;**31**(1): 2181959. DOI: 10.1080/27690911.2023.2181959

[7] Shams M, Rafiq N, Kausar N, Agarwal P, Mir NA, Li YM. On highly efficient simultaneous schemes for finding all polynomial roots. Fractals. 2022;**30**(10):2240198. DOI: 10.1142/S0218348X22401983

[8] Bland DR. Solutions of Laplace's Equation. London: Routledge and Keegan Paul Ltd.; 1961

[9] Hayt W, Buck J. Engineering Electromagnetics. New York: The McGraw-Hill; 2012

[10] Jackson JD. Classical Electrodynamics. John Wiley Sons; 2021

[11] Kreyszig E. Advanced Engineering Mathematics 9th Edition with Wiley Plus Set. US: John Wiley Sons; 2007

[12] Zill DG. A First Course in Differential Equations. Bostonn MA: PWS-Kent; 1993

[13] Phillips GM. Interpolation and Approximation by Polynomials. New York: Springer Science Business Media; 2003

[14] Erturk VS, Momani S, Odibat Z. Application of generalized differential transform method to multi-order fractional differential equations. Communications in Nonlinear Science and Numerical Simulation. 2008;**13**: 1642-1654. DOI: 10.1016/j.cnsns.2007.02.006

[15] Odibat Z, Momani S. A generalized differential transform method for linear partial differential equations of fractional order. Applied Mathematics Letters. 2008;**21**(2):194-199. DOI: 10.1016/j.aml.2007.02.022

[16] Najafi HS, Mirshafaei SR, Toroqi EA. An approximate solution of the Mathieu fractional equation by using the generalized differential transform method (GDTM). Applications and Applied Mathematics: An International Journal (AAM). 2012;**7**(1):24. Available

from: https://digitalcommons.pvamu.edu/aam/vol7/iss1/24

[17] Najafi HS, Toroqi EA, Divishali AJ. A new fractional model of single degree of freedom system, by using generalized differential transform method. Romanian Journal of Mathematics and Computer Science. 2016;**6**(1):93-105. Available from: https://www.academia.edu/download/77688811/Najafi-Toroqi-Divishali-2016.pdf

[18] Najafi HS, Toroqi EA. Generalized differential transform method for solving liquid-film mass transfer equation of fractional order. Progress in Fractional Differentiation and Applications. 2018;**4**:59-64. Available from: https://www.naturalspublishing.com/download.asp?ArtcID=12817

[19] Fitzpatrick R. Maxwell's Equations and the Principles of Electromagnetism. Hingham, Massachusetts: Jones Bartlett Publishers; 2008

[20] Bakshi AV, Bakshi UV. Electromagnetic Theory. Pune, India: Technical Publications; 2008

[21] Thomas JW. Numerical Partial Differential Equations: Finite Difference Methods. New York: Springer Science Business Media; 2013. DOI: 10.1007/978-1-4899-7278-1

[22] Vanderlinde J. The Covariant Formulation. Classical Electromagnetic Theory. 2005;**145**:313-329. DOI: 10.1007/1-4020-2700-1

[23] Odibat ZM, Shawagfeh NT. Generalized Taylor's formula. Applied Mathematics and Computation. 2007; **186**(1):286-293. DOI: 10.1016/j.amc.2006.07.102

[24] Odibat Z, Momani S, Erturk VS. Generalized differential transform method: Application to differential equations of fractional order. Applied Mathematics and Computation. 2008; **197**(2):467-477. DOI: 10.1016/j.amc.2007.07.068

[25] Fehlberg E. Classical Seventh-, Sixth-, and Fifth-Order Runge-Kutta-Nystróm Formulas with Stepsize Control for General Second-Order Differential Equations; 1974. Document ID 19740026877

[26] Fehlberg E. Klassische Runge-Kutta-Formeln vierter und niedrigerer Ordnung mit Schrittweiten-Kontrolle und ihre Anwendung auf Warmeleitungs-Probleme. Computing. 1970;**6**:61-71

Chapter 6

Numerical Solution of the Equation with an Angular Boundary Layer

Asan Sydygalievich Omuraliev, Peyil Esengul Kyzy and Ella Dayrbekovna Abylaeva

Abstract

This work is devoted to a numerical method for solving a first-order partial differential equation when a small parameter is present in all derivatives. A numerical method based on a synthesis of S.A. The Lomov method and the finite-difference method are proposed. In the second part, a regularized asymptotic of any order of a mixed problem is constructed for a first-order partial differential equation, when the limit equation has a regular singularity. The constructed asymptotic contains boundary layer functions of three types: power, exponential, and angular functions. The asymptotic of the solution is constructed by a special class of functions corresponding to the structure of the fundamental system of solutions. The solution to the problem is sought in the form of a polynomial in powers of a small parameter. The asymptotic character of the constructed solution is established.

Keywords: power-law boundary layer, angular boundary layer, singular perturbation, regularization, asymptotics, numerical solution

1. Introduction

Mathematical modeling in the natural and technical sciences has stimulated the development of numerous algorithms for the approximate integration of differential equations. Using such algorithms, it is easy to calculate approximate solutions to systems with dozens of equations if the latter satisfy the conditions of the existence theorem, uniqueness, and continuous dependence of solutions on a parameter. In cases where the conditions of the said theorem are violated, the efficiency of the algorithms is significantly reduced and, in some situations, becomes practically zero. The difficulties that arise can be overcome using a preliminary asymptotic analysis of the problem under study, carried out on the basis of methods of asymptotic integration of differential equations. The combination of numerical integration algorithms with asymptotic integration methods allows us to develop an algorithm for the approximate solution of differential equations, significantly reducing the number of calculations and making the computational process most expedient and optimal. Currently, there are a sufficient number of numerical methods for solving stiff problems. Numerical solutions to such problems are devoted to Refs. [1–25]. In many of the works cited on numerical solutions of singularly perturbed problems, the so-called adjustable parameter [26] is used in the discretization of the original problem. It contains an irregular dependence on a small parameter, which is included there in the form of an exponential function. These methods allow for constructing solutions with an accuracy of the order of h or h^2.

An analysis of works on numerical algorithms for solving stiff problems shows that the constructed algorithms are stable if the obtained solution contains a boundary layer-type function. A boundary layer-type function can be included in the solution of the problem both linearly and nonlinearly.

So far, such problems have not been studied from the standpoint by the method of singularly perturbed problems. In the works of Vasilyeva et al. [27–30], the asymptotic of the boundary layer type was constructed. The asymptotic constructed there has a complex structure and the process of constructing the solution consists of several stages. In the work of Vasilyeva [27], for a system of equations in partial derivatives of the first order, the asymptotic of the inner transitional layer was constructed. In the works of Nesterov et al. [28–30], the Cauchy problems for systems of singularly perturbed partial differential equations of the first order are studied, when the matrix at the desired function has one zero eigenvalue, and an asymptotic of any order of the boundary layer type is constructed. Our approach greatly simplifies the process of constructing the asymptotic.

The numerical solution of singularly perturbed problems is the subject of works whose bibliography is given in the works [31, 32]. The works [31–36] are devoted to the construction of difference schemes for singularly perturbed ordinary differential equations. In the works [33–35, 37], singularly perturbed problems were studied on a uniform piecewise grid. The works [38–40] use the decomposition of the grid solution into regular and singular components, which are solutions of grid subproblems on piecewise uniform grids.

This work proposes a new approach for the numerical solution of singularly perturbed ordinary differential equations, which is based on the synthesis of S.A. Lomov's regularization method [41] and known numerical methods (finite elements, finite differences, direct lines). The idea of the method is to regularize a singularly perturbed problem, by introducing an additional regularizing independent variable, the original problem is expanded into a space of higher dimension. The extended problem obtained in this case will be regular in a small parameter, then the resulting regular problem is decomposed, the resulting equations for the components are applied, and one of the known numerical methods is applied. Previously, this method was applied in the works [42–46] to various singularly perturbed ordinary differential equations, and in the work of Omuraliev [47] to a parabolic equation. The method of lines in the work of Omuraliev [42] solved the initial problem for a differential equation with a small first-order parameter. Singularly perturbed ordinary differential equations with one boundary layer function and two boundary layer functions, based on the finite difference method, were studied, respectively, in the works of Omuraliev [43, 44]. The finite element method was applied in the works of Omuraliev [45, 46] to solve singularly perturbed ordinary differential equations. In the work of Omuraliev [47], the finite difference method is used to solve a singularly perturbed heat equation.

2. The equation with an angular boundary layer

The problem under study is the solution of which contains a discontinuous angular boundary layer function.

Let us consider the problem

$$L_\varepsilon u \equiv \varepsilon(\partial_t u(x,t,\varepsilon) + a(x,t)\partial_x u(x,t,\varepsilon)) + b(x,t)u(x,t,\varepsilon) = f(x,t),$$
$$u_{t=0} = \alpha(x),\ \ u_{x=0} = \beta(t),\ \ (x,t)\in\Omega = (0,1]\times(0,T], \qquad (1)$$

under the following assumptions:
1) $a(x,t), b(x,t) > 0\ \ \forall(x,t)\in\overline{\Omega}$;
2) $a(x,t), b(x,t), f(x,t)\in C(\overline{\Omega})$, $\alpha(x)\in C[0,1]$, $\beta(t)\in C[0,T]$.

2.1 Regularization of the problem

We introduce regularizing variables

$$\eta = \frac{\varphi(x,t)}{\varepsilon}, \varphi(x,0) = 0,\quad \xi = \frac{\psi(x,t)}{\varepsilon}, \psi(0,t) = 0 \qquad (2)$$

and an extended function $\tilde{u}(M,\varepsilon)$ such that

$$\tilde{u}(M,\varepsilon)_{\gamma=\theta(x,t,\varepsilon)} \equiv u(x,t,\varepsilon), \qquad (3)$$
$$\gamma = (\eta,\xi),\ \ \theta(x,t,\varepsilon) = \left(\frac{\varphi(x,t)}{\varepsilon}, \frac{\psi(x,t)}{\varepsilon}\right),\ \ M = (x,t,\eta,\xi).$$

From (3), based on (2), we find the derivatives with respect to x and t, then instead of problem (1), we pose the extended problem

$$\tilde{L}_\varepsilon \tilde{u} \equiv \varepsilon(\partial_t\tilde{u} + a(x,t)\partial_x\tilde{u}) + (\partial_t\varphi + a(x,t)\partial_x\varphi)\partial_\eta\tilde{u} +$$
$$+(\partial_t\psi + a(x,t)\partial_x\psi)\partial_\xi\tilde{u} + b(x,t)\tilde{u} = f(x,t),\ \ M\in Q \qquad (4)$$
$$\tilde{u}(M,\varepsilon)_{t=\eta=0} = \alpha(x),\ \ \tilde{u}(M,\varepsilon)_{x=\xi=0} = \beta(t),\ \ Q = \Omega\times(0,\infty)^2.$$

Regularizing functions are selected as solutions to problems

$$\partial_t\varphi + a(x,t)\partial_x\varphi = b(x,t), \varphi(x,0) = 0,\ \ \partial_t\psi + a(x,t)\partial_x\psi = b(x,t), \psi(0,t) = 0. \qquad (5)$$

The solution to problem (4) is defined as

$$\tilde{u}(M,\varepsilon) = v_1(x,t) + v_2(x,t) + c_1(x,t)\exp(-\eta) + c_2(x,t)\exp(-\xi) +$$
$$\omega_1(x,t)\Phi(\eta-\xi)\exp(-\xi) + \omega_2(x,t)\Phi(\xi-\eta)\exp(-\eta), \qquad (6)$$

where

$$\Phi(z) = \begin{cases} 0, & if\ \ z<0 \\ \exp(-z), & if\ \ z\geq 0. \end{cases}$$

Substituting (6) into problem (4), taking into account (5), regarding the components we obtain the following problems:

$$b(x,t)v_1(x,t) = f(x,t),$$

$$\varepsilon(\partial_t v_2(x,t) + a(x,t)\partial_x v_2(x,t)) + b(x,t)v_2(x,t) = -\varepsilon(\partial_t v_1(x,t) + a(x,t)\partial_x v_1(x,t)),$$
$$v_2(x,0) = 0,$$
$$\partial_t c_1(x,t) + a(x,t)\partial_x c_1(x,t) = 0, \;\; c_1(x,0) = \alpha(x) - v_1(x,0), \tag{7}$$
$$\partial_t c_2(x,t) + a(x,t)\partial_x c_2(x,t) = 0, \;\; c_2(0,t) = \beta(t) - v_1(0,t),$$
$$\partial_t \omega_1(x,t) + a(x,t)\partial_x \omega_1(x,t) = 0, \;\; \omega_1(0,t) = -c_1(0,t),$$
$$\partial_t \omega_2(x,t) + a(x,t)\partial_x \omega_2(x,t) = 0, \;\; \omega_2(x,0) = -c_2(x,0).$$

2.2 Numerical solution

We will approximate continuous problems (7) on a uniform grid $x_i = (i-1)h, i = 1,2,\dots,n+1,\;\; n = 1/h,\;\; t_j = (j-1)\tau,\;\; j = 1,2,\dots,m+1,\;\; m = T/\tau.$ Using finite differences

$$\partial_t u = \frac{u_i^{j+1} - u_i^j}{\tau} + O(\tau)$$

$$\partial_x u = \frac{u_{i+1}^j - u_i^j}{h} + O(h)$$

of problem (7), we replace them with difference schemes:

$$v_{1,i}^j = \frac{f_i^j}{b_i^j}, \;\; i = 1,2,\dots,n+1, \;\; j = 1,2,\dots,m+1, \tag{8}$$

$$\varepsilon\left(\frac{v_{2,i}^{j+1} - v_{2,i}^j}{\tau} + a_i^j \frac{v_{2,i+1}^j - v_{2,i}^j}{h}\right) + b_i^j v_{2,i}^j = f_{1,i}^j + O(\varepsilon(\tau + h)), \tag{9}$$

$$v_{2,i}^1 = 0, \;\; v_{2,1}^j = 0, \;\; f_{1,i}^j = -\varepsilon\left(\frac{v_{1,i}^{j+1} - v_{1,i}^j}{\tau} + a_i^j \frac{v_{1,i+1}^j - v_{1,i}^j}{h}\right),$$

$$\frac{c_{1,i}^{j+1} - c_{1,i}^j}{\tau} + a_i^j \frac{c_{1,i+1}^j - c_{1,i}^j}{h} = O(\tau + h), \;\; c_{1,i}^1 = \alpha_i - v_{1,i}^1, \tag{10}$$

$$\frac{c_{2,i}^{j+1} - c_{2,i}^j}{\tau} + a_i^j \frac{c_{2,i+1}^j - c_{2,i}^j}{h} = O(\tau + h), \;\; c_{2,1}^j = \beta_j - v_{1,1}^j, \tag{11}$$

$$\frac{\omega_{1,i}^{j+1} - \omega_{1,i}^j}{\tau} + a_i^j \frac{\omega_{1,i+1}^j - \omega_{1,i}^j}{h} = O(\tau + h), \;\; \omega_{1,1}^j = -c_{1,1}^j, \tag{12}$$

$$\frac{\omega_{2,i}^{j+1} - \omega_{2,i}^j}{\tau} + a_i^j \frac{\omega_{2,i+1}^j - \omega_{2,i}^j}{h} = O(\tau + h), \;\; \omega_{2,i}^1 = -c_{2,i}^1, \tag{13}$$

for the regularizing functions and obtain the discrete problems:

$$\varphi_{i+1}^j = \varphi_i^j - \frac{h}{\tau a_i^j}\left(\varphi_i^{j+1} - \varphi_i^j\right) + \frac{h}{a_i^j} b_i^j, \;\; \varphi_1^j = 0, \tag{14}$$

$$\psi_i^{j+1} = \psi_i^j - \frac{\tau a_i^j}{h}\left(\psi_{i+1}^j - \psi_i^j\right) + \tau b_i^j, \;\; \psi_i^1 = 0. \tag{15}$$

Problems (8)–(15) are explicit difference schemes, from which we will determine the solutions of these problems and substitute them into (6), having previously put there

$$x = x_i,\ t = t_j,\ \xi_{1,i} = \frac{1}{\varepsilon}\varphi_i^j,\ \xi_{2,i} = \frac{1}{\varepsilon}\psi_i^j,$$
$$v_l(x_i, t_j) = v_{l,i}^j,\ c_l(x_i, t_j) = c_{l,i}^j,\ \omega_l(x_i, t_j) = \omega_{l,i}^j \tag{16}$$

This will give us a numerical solution to the original problem (1). Following work ([26] pp. 346–348) to solve problem (9), we obtain the estimate

$$\|v_2\| < c(\varepsilon(\tau + h)) \tag{17}$$

and proceeding in the same way as work ([26] pp. 333–334), we will obtain estimates for the solution of problems (10)–(12):

$$\|c_l\|, \|O_l\| < c(\tau + h),\ \ l = 1, 2.$$

Based on these estimates from (6), on the basis of (16) and the fact that the functions c_l and O_l are multiplied by exponentially small functions, to solve the original problem we obtain an estimate of the form (17).

Example: Consider the problem

$$\varepsilon(\partial_t u + \partial_x u) + bu = (1 + x)\exp(-t),\ u_{t=0} = x,\ u_{x=0} = t$$

the solution algorithm is as follows

$$b(\partial_\tau \tilde{u} + \partial_\xi \tilde{u} + b\tilde{u}) + \varepsilon(\partial_t \tilde{u} + \partial_x \tilde{u}) = (1 + x)e^{-t},$$
$$\tilde{u}_{\tau=t=0} = x,\ \tilde{u}_{\xi=x=0} = t,\ \xi = \frac{x}{\varepsilon},\ \tau = \frac{t}{\varepsilon}.$$

The exact solution to this problem can be represented as:

$$ut = \beta(t - x)\exp\left(-\frac{bx}{\varepsilon}\right) - \frac{(1 + x - t)}{\varepsilon - b}\exp(-t)\left(1 - \exp\left(\frac{(\varepsilon - b)x}{\varepsilon}\right)\right) -$$
$$-\frac{\exp(-t)}{\varepsilon - b}\left(t - (t - x)\exp\left(\frac{(\varepsilon - b)x}{\varepsilon}\right)\right) - \frac{\varepsilon\exp(-t)}{(\varepsilon - b)^2}\left(1 - \exp\left(\frac{(\varepsilon - b)x}{\varepsilon}\right)\right) =$$
$$\alpha(x - t)\exp\left(-\frac{bt}{\varepsilon}\right) - \frac{1 + x - t}{\varepsilon - b}\exp(-t)\left(1 - \exp\left(-\frac{bt}{\varepsilon}\right)\right) - \frac{t}{\varepsilon - b}\exp(-t) -$$
$$-\frac{\varepsilon}{(\varepsilon - b)^2}\left(\exp(-t) - \exp\left(-\frac{bt}{\varepsilon}\right)\right),$$
$$\alpha(x - t) = \begin{cases} x - t, & \textit{if } x - t > 0 \\ 0, & \textit{if } x - t < 0. \end{cases}$$
$$\beta(t - x) = \begin{cases} t - x, & \textit{if } t - x > 0 \\ 0, & \textit{if } t - x < 0. \end{cases}$$

(a)										
					ut=					
0	0.1000	0.2000	0.3000	0.4000	0.5000	0.6000	0.7000	0.8000	0.9000	1.0000
0.1000	0.3322	0.3002	0.2716	0.2458	0.2224	0.2012	0.1821	0.1648	0.1491	0.1350
0.2000	0.3619	0.3275	0.2963	0.2681	0.2426	0.2195	0.1986	0.1797	0.1626	0.1472
0.3000	0.3921	0.3548	0.3210	0.2905	0.2628	0.2378	0.2152	0.1947	0.1762	0.1594
0.4000	0.4223	0.3821	0.3457	0.3128	0.2830	0.2561	0.2317	0.2097	0.1897	0.1717
0.5000	0.4524	0.4094	0.3704	0.3352	0.3033	0.2744	0.2483	0.2247	0.2033	0.1839
0.6000	0.4826	0.4367	0.3951	0.3575	0.3235	0.2927	0.2648	0.2396	0.2168	0.1962
0.7000	0.5127	0.4639	0.4198	0.3798	0.3437	0.3110	0.2814	0.2546	0.2304	0.2085
0.8000	0.5429	0.4912	0.4445	0.4022	0.3639	0.3293	0.2980	0.2696	0.2439	0.2207
0.9000	0.5731	0.5185	0.4692	0.4245	0.3841	0.3476	0.3145	0.2846	0.2575	0.2330
1.0000	0.6033	0.5458	0.4939	0.4469	0.4044	0.3659	0.3311	0.2996	0.2710	0.2453
(b)										
					u=					
0	0.1000	0.2000	0.3000	0.4000	0.5000	0.6000	0.7000	0.8000	0.9000	1.0000
0.1000	0.3320	0.3004	0.2719	0.2460	0.2226	0.2014	0.1823	0.1649	0.1492	0.1350
0.2000	0.3625	0.3280	0.2968	0.2686	0.2430	0.2199	0.1990	0.1800	0.1629	0.1474
0.3000	0.3930	0.3556	0.3217	0.2911	0.2634	0.2384	0.2157	0.1951	0.1766	0.1598
0.4000	0.4234	0.3831	0.3467	0.3137	0.2838	0.2568	0.2324	0.2103	0.1903	0.1722
0.5000	0.4539	0.4107	0.3716	0.3363	0.3043	0.2753	0.2491	0.2254	0.2040	0.1845
0.6000	0.4844	0.4383	0.3966	0.3588	0.3247	0.2938	0.2658	0.2405	0.2176	0.1969
0.7000	0.5148	0.4658	0.4215	0.3814	0.3451	0.3123	0.2826	0.2557	0.2313	0.2093
0.8000	0.5453	0.4934	0.4465	0.4040	0.3655	0.3307	0.2993	0.2708	0.2450	0.2217
0.9000	0.5758	0.5210	0.4714	0.4265	0.3860	0.3492	0.3160	0.2859	0.2587	0.2341
1.0000	0.6063	0.5485	0.4963	0.4491	0.4064	0.3677	0.3327	0.3011	0.2724	0.2465

Table 1.
(a) ut is exact solution (b) u is numerical solution.

The solution to the extended problem is defined as

$$\tilde{u} = v_1(x,t) + v_2(x,t) + c_1(x,t)\exp(-\tau) + c_2(x,t)\exp(-\xi) + \\ + \Phi(\tau-\xi)\omega_1(x,t)\exp(-\xi) + \Phi(\xi-\tau)\omega_2(x,t)\exp(-\tau),$$

The components of this function are determined from the following difference schemes

$$v_1(i,j) = (1 + x(i))\exp(-t(j)),$$

$$v_2(i,j+1) = \left(1 + \frac{\tau}{h} - \tau b\right)v_2(i,j) - \frac{\tau}{h}v_2(i+1,j) - (v_1(i,j+1) - v_1(i,j))$$

$$v_2(i,1) = 0, \ \ v_2(1,j) = 0.$$

$$c_1(i,j+1) = \left(1+\frac{\tau}{h}\right)c_1(i,j) - \frac{\tau}{h}c_1(i+1,j), \ \ c_1(i,1) = x(i) - v_1(i,1),$$

$$c_2(i,j+1) = \left(1+\frac{h}{\tau}\right)c_2(i,j) - \frac{h}{\tau}c_2(i+1,j), \ \ c_2(i,1) = t(j) - v_1(i,1),$$

$$\omega_1(i+1,j) = \left(1+\frac{\tau}{h}\right)\omega_1(i,j) - \frac{\tau}{h}\omega_1(i,j+1), \ \ \omega_1(1,j) = -c_1(1,j),$$

$$\omega_2(i,j+1) = \left(1+\frac{h}{\tau}\right)\omega_2(i,j) - \frac{\tau}{h}\omega_2(i+1,j), \ \ \omega_2(i,1) = c_2(i,1).$$

The numerical solution was obtained in the MATLAB package for $\varepsilon = 0.03, \ h = 0.1, \ \tau = 0.1$. Here ut is the exact solution, and u is the numerical solution (**Table 1**).

3. The equation with angular and power-law boundary layer

3.1 Asymptotic solution

Consider the problem

$$(\varepsilon + t)\partial_t u + \varepsilon a(x)\partial_x u + b(x,t)u = f(x,t), \ \ (x,t) \in \Omega$$
$$u(x,t,\varepsilon)\big|_{t=0} = u^0(x), \ \ u(x,t,\varepsilon)\big|_{x=0} = u^1(t) \tag{18}$$

here $\varepsilon > 0$ is a small parameter, $a(x) \in C^\infty[0,1], \ b(t) \in C^\infty(\Omega), \ f(x,t) \in C^\infty(\Omega), \ \Omega = \{0 < x < 1, 0 < t \le T\}$.

The problem is studied at $b(x,0) > 0, \ \forall t \in [0,T]$. The degenerate $(\varepsilon = 0)$ equation has a singularity at $t = 0$, which leads to the appearance of a power-law boundary layer. The power-law boundary layer [27] is described by the function

$$\Pi(t,\varepsilon) = \left(\frac{\varepsilon}{t+\varepsilon}\right)^\lambda, \ \ \lambda > 0,$$

in addition, the problem (18) along the characteristic has a gap.

We have constructed a continuous asymptotic solution that contains regular, power, and angular boundary layer functions. Previously, the problem solutions of which contain power-law boundary layer functions were studied in the works [41, 48–51]. In the works [41, 48] are studied ordinary differential equations, and works [49–51] are devoted to the construction of an asymptotic solution of parabolic equations.

3.2 Regularization of the problem

Let us regularize [41] the problem (18), for which we introduce the regularizing functions

$$\xi_1 = \varphi_1(x,t,\varepsilon), \ \ \xi_2 = \varphi_2(x,t,\varepsilon), \ \ \varphi_1(x,0,\varepsilon) = 0, \ \ \varphi_2(0,t,\varepsilon) = 0 \tag{19}$$

and the extended function

$$\tilde{u}(x,t,\xi,\varepsilon)|_{\xi=\varphi(x,t,\varepsilon)}u(x,t,\varepsilon),\ \ \xi=(\xi_1,\xi_2),\ \ \varphi=(\varphi_1,\varphi_2). \tag{20}$$

From (20), based on (19), we find the derivatives of $\partial_t u,\ \ \partial_x u$ and choose the regularizing functions as solutions of the equations

$$\varepsilon a(x)\partial_x\varphi_2(x,t,\varepsilon)+(\varepsilon+t)\partial_t\varphi_2(x,t,\varepsilon)=b(x,0),\ (\varepsilon+t)\partial_t\varphi_1(x,t,\varepsilon)+\varepsilon a(x)\partial_x\varphi_1(x,t,\varepsilon)=b(x,0) \tag{21}$$

Then the extended task for $\tilde{u}(x,t,\xi,\varepsilon)$ will be written as

$$b(x,0)\partial_\tau\tilde{u}+b(x,0)\partial_\xi\tilde{u}+b(x,0)\tilde{u}+[b(x,t)-b(x,0)]\tilde{u}+t\partial_t\tilde{u}=-\varepsilon\partial_t\tilde{u}-\varepsilon a(x)\partial_x\tilde{u}+f(x,t),\quad (x,t,\xi)\in Q, \tag{22}$$

$$\tilde{u}|_{t=\tau=0}=u^0(x),\ \ \tilde{u}\big|_{x=\xi=0}=u^1(t),\ \ Q=\Omega\times(0,\infty)\times(0,\infty)$$

Solving problems (21), (19) will be written as:

$$\varphi_1(x,t,\varepsilon)=\int_0^\tau b\left(A^{-1}(\varepsilon\eta-\tau+s),0\right)ds,\ \ \tau=b(x,0)\,ln\left(\frac{t+\varepsilon}{\varepsilon}\right), \tag{23}$$

$$\varphi_2(x,t,\varepsilon)=\int_0^\eta b\left(A^{-1}(\varepsilon s),0\right)ds,\ \ \eta=\frac{1}{\varepsilon}\int_0^x\frac{ds}{a(s)}\equiv\frac{1}{\varepsilon}A(x)$$

The solution of problem (22) will be defined as a series

$$\tilde{u}(x,t,\xi,\tau,\varepsilon)=\sum_{k=0}^{\infty}\varepsilon^k u_k(M),\ \ M=(x,t,\xi,\tau),$$

then for the coefficients, we get the following iterative problems:

$$\begin{gathered}Tu_0\equiv b(x,0)[\partial_\tau+\partial_\xi+1]u_0+t\partial_t u_0+[b(x,t)-b(x,0)]u_0=f(x,t),\\ Tu_k=-\partial_t u_{k-1}-a(x)\partial_x u_{k-1},\\ u_0|_{t=\tau=0}=u^0(x),\ \ u_0\big|_{x=\xi=0}=u^1(t),\\ u_k|_{t=\tau=0}=u_k|_{x=\xi=0}=0.\end{gathered} \tag{24}$$

3.3 Solvability of iterative problems

Iterative problems (24) will be solved in the class of functions

$$\begin{gathered}U=\{u(M):u(M)=c_1(x,t)e^{-\tau}+d_1(x,t)e^{-\xi}+e_1(x,t)\Phi(\tau-\xi)e^{-\xi}+\\ f_1(x,t)\Phi(\xi-\tau)e^{-\tau}+v(x,t),\\ c_1(x,t),d_1(x,t),e_1(x,t),f_1(x,t),v(x,t)\in C^\infty\left(\overline{\Omega}\right)\}\end{gathered}$$

$$\Phi(\xi)=\begin{cases}0, & \xi<0,\\ e^{-\xi}, & \xi\geq 0.\end{cases}$$

where the term $c_1(x,t)e^{-\tau} = c_1(x,t)\left(\frac{\varepsilon}{t+\varepsilon}\right)^{b(x,0)}$ describes a power boundary layer along $t = 0, d_1(x,t)e^{-\xi}$ describes an exponential boundary layer along $x = 0$; the remaining two terms describe an angular boundary layer in the vicinity of point $(0, 0)$.

Calculate the action of the operator T on the function $u(M) \in U$:

$$\begin{gathered} Tu = b(x,0)\left[-c_1(x,t)e^{-\tau} - d_1(x,t)e^{-\xi} - e_1(x,t)\Phi(\tau-\xi)e^{-\xi} - \right. \\ f_1(x,t)\Phi(\xi-\tau)e^{-\tau} + e_1(x,t)\Phi'(\tau-\xi)e^{-\xi} - e_1(x,t)\Phi'(\tau-\xi)e^{-\xi} + \\ f_1(x,t)\Phi'(\xi-\tau)e^{-\tau} - f_1(x,t)\Phi'(\xi-\tau)e^{-\tau} + \\ c_1(x,t)e^{-\tau} + d_1(x,t)e^{-\xi} + e_1(x,t)\Phi(\tau-\xi)e^{-\xi} + \\ \left. f_1(x,t)\Phi(\xi-\tau)e^{-\tau}\right] + t\left[\partial_t c_1(x,t)e^{-\tau} + \partial_t d_1(x,t)e^{-\xi} + \right. \\ \left. \partial_t e_1(x,t)\Phi(\tau-\xi)e^{-\xi} + \partial_t f_1(x,t)\Phi(\xi-\tau)e^{-\tau}\right] + \\ [b(x,t) - b(x,0)]\left[c_1(x,t)e^{-t} + d_1(x,t)e^{-\xi} + e_1(x,t)\Phi(\tau-\xi)e^{-\xi} + \right. \\ \left. f_1(x,t)\Phi(\xi-\tau)e^{-\tau}\right] = [t[\partial_t c_1(x,t) + (b(x,t) - b(x,0))c_1(x,t)]e^{-\tau} + \\ [t\partial_t d_1(x,t) + (b(x,t) - b(x,0))d_1(x,t)]e^{-\xi} + \\ [t\partial_t e_1(x,t) + (b(x,t) - b(x,0))e_1(x,t)] \times \Phi(\tau-\xi)e^{-\xi} + \\ \left[t\partial_t f_1(x,t) - (b(x,t) - b(x,0))f_1(x,t)\right]\Phi(\xi-\tau)e^{-\tau} + \\ b(x,t)v(x,t) + t\partial_t v(x,t), \end{gathered} \tag{25}$$

here

$$\begin{aligned} c_1(x,t) &= c(x,t) + P_1(x), \\ d_1(x,t) &= d(x,t) + P_2(x), \\ f_1(x,t) &= f(x,t) + P_4(x), \\ e_1(x,t) &= e(x,t) + P_3(x). \end{aligned}$$

From the boundary conditions (24) of the function $u(M) \in U$, we find

$$\begin{aligned} c_1(x,t) &= -v(x,0) - P_1(x), \quad f_1(x,t) = -d_1(x,t) - P_4(x), \\ d_1(0,t) &= -v(0,t) - P_2(0), \quad e_1(0,t) = -c_1(0,t) - P_3(0). \end{aligned} \tag{26}$$

Satisfying the function $u_0(M) \in U$ to Eq. (24) for $k = 0$, based on calculations (25), we obtain

$$\begin{aligned} t\partial_t c^0(x,t) + [b(x,t) - b(x,0)]\left[c^0(x,t) - P_1^0(x)\right] &= 0, \\ t\partial_t d^0(x,t) + [b(x,t) - b(x,0)]\left[d^0(x,t) - P_2^0(x)\right] &= 0, \\ t\partial_t e^0(x,t) + [b(x,t) - b(x,0)]\left[e^0(x,t) - P_3^0(x)\right] &= 0, \\ t\partial_t f^0(x,t) + [b(x,t) - b(x,0)]\left[f^0(x,t) - P_4^0(x)\right] &= 0. \end{aligned} \tag{27}$$

These equations, under initial conditions (26), have smooth solutions. The functions $P_i^0(x), \quad i = 1,2,3,4$ included here and in the initial condition will be defined in the next iteration step. In the next iteration step, the right side of the equations will include

$$F_1(M) = -\partial_t u_0 - a(x)\partial_x u_0.$$

We substitute only the term $\left(c^0(x,t) + P_1^0(x)\right)e^{-\tau}$, into it, the other terms of the function $u_0(M)$ are transformed in the same way:

$$F_2(M) = -\left[\partial_t c^0(x,t) + a(x)\left(\partial_x c^0(x,t) + \frac{dP_1^0}{dx}\right)\right]e^{-\tau}.$$

To ensure the solvability of the equation with respect to $c^1(x,t)$ from (27), we assume

$$\frac{dP_1^0(x)}{dx} = -\left(\partial_t c^0(x,t) + a(x)\partial_x c^0(x,t)\right)\frac{1}{a(x)}\bigg|_{t=0}.$$

Substituting here the value of $c^0(x,t)$ found as the solution of problem (27), (26), with respect to $P_1^0(x)$, we obtain the equation. The resulting equation is solved under an arbitrary initial condition for $x = 0$.

Then the process repeats. The asymptotic character of the constructed solution is proved.

Theorem 1.1 The given functions satisfy the following conditions:
$a(x) \in C^\infty([0,1]), b(x,t), f(x,t) \in C^\infty(\overline{\Omega})$ and the initial conditions. Then, for sufficiently small $\varepsilon > 0$, problem (18) has a smooth asymptotic solution, i.e. there is an estimate

$$|u(x,t,\varepsilon) - u_{\varepsilon_n}(x,t,\xi,\varepsilon)| < c\varepsilon^{n+1}, \quad \forall n \geq 0.$$

Proof: Let us rewrite problem (18)

$$\partial_t u + \frac{\varepsilon}{\varepsilon + t}a(x)\partial_x u + \frac{1}{\varepsilon + t}b(x,t)u = \frac{1}{\varepsilon + t}f(x,t).$$

Here, the expression $(t + \varepsilon)$ for sufficiently small ε does not affect the properties of the function $a(x), b(x,t)$ for which the conditions of the maximum principle theorem are valid [52]. Therefore, on the basis of this theorem, it is not difficult to establish an estimate.

3.4 Numerical solution

Our method is based on the method of S. A. Lomov. First, the singularly perturbed equation under study is reduced by these methods to a regularly perturbed equation, then the resulting equation is decomposed. The equations obtained after decomposition with initial conditions are solved by a well-known numerical method. In the numerical solution in Eq. (21), $b(x,-\varepsilon)$ is taken instead of $b(x,0)$.

In this work, the finite difference method is used. The solution of problem (22) will be defined as

$$\tilde{u}(x,t,\xi,\tau,\varepsilon) = c^1(x,t)e^{-\tau} + c^2(x,t)e^{-\xi} + \Phi(\xi-\tau)c^3(x,t)e^{-\tau} +$$
$$+c^4(x,t)\Phi(\tau-\xi)e^{-\xi} + v(x,t), \tag{28}$$

for the coefficients, we obtain the problem

$$(\varepsilon+t)\partial_t c^l(x,t) + \varepsilon a(x)\partial_x c^l(x,t) +$$
$$+[b(x,t) - b(x,-\varepsilon)]c^l(x,t) = 0, l = 1,2,3,4, \tag{29}$$
$$(\varepsilon+t)\partial_t v(x,t) + \varepsilon a(x)\partial_x v(x,t) + b(x,t)v(x,t) = f(x,t) \tag{30}$$

Eq. (30) is solved without the initial condition, and for eqs. (29), the initial conditions are given in the form

$$c^1(x,0) = u^0(x) - v(x,0),\ \ c^3(x,0) = -c^2(x,0),$$
$$c^2(0,t) = u^1(t) - v(0,t),\ \ c^4(0,t) = -c^1(0,t). \tag{31}$$

Difference equations equivalent to these problems can be written as

$$(\varepsilon+t_j)\frac{c^l_{i,j+1} - c^l_{i,j}}{k} + \varepsilon a_i \frac{c^l_{i+1,j} - c^l_{i,j}}{h} + \left[b_{ij} - b_i\right]c^l_{ij} = O(k+h),\ \ l = \overline{1},4 \tag{32}$$

$$(\varepsilon+t_j)\frac{v_{i,j+1} - v_{i,j}}{k} + \varepsilon a_i \frac{v_{i+1,j} - v_{i,j}}{h} + b_{ij}v_{ij} = f_{ij} + O(k+h),\ \ i = \overline{1},n, j = \overline{1},m \tag{33}$$

$$h = \frac{1}{n}, k = \frac{1}{m},\ \ c^1_{i,0} = u^0_i - v_{i,0},\ \ c^3_{i,0} = -c^2_i,\ \ c^2_{0,j} = u^1_j - v_{0,j},$$
$$c^4_{o,j} = -c^1_{0,j},\ \ b_{ij} = b(ih, jl), b_i = b(ih, -\varepsilon).$$

Eq. (32) with $l = 1, 3$, we write

$$c^l_{i,j+1} = q_{ij}c^l_{ij} - p_{ij}c^l_{i+1,j} + O(k+h),\ \ c^1_{i,0} = u^0_i - v_{i,0},\ \ c^3_{i,0} = -c^2_{i,0}, \tag{34}$$
$$q_{ij} = 1 + \frac{\varepsilon a_i r - (b_{ij} - b_i)k}{\varepsilon + t_j},\ \ r = \frac{k}{h},\ \ p_{ij} = \frac{\varepsilon a_i}{t_j + \varepsilon}r,$$

for $l = 2, 4$, we write

$$c^l_{i+1,j} = q^1_{ij}c^l_{ij} - r_1 p^1_{ij}c^l_{i,j+1} + O(k+h),\ \ c^2_{0,j} = u^1_j - v_{0,j},\ \ c^4_{0,j} = -c^1_{oj}, \tag{35}$$
$$q^1_{ij} = 1 + \frac{r_1(\varepsilon + t_j) - (b_{ij} - b_i)h}{\varepsilon a_i},\ \ p^1_{ij} = \frac{r_1(t_j + \varepsilon)}{\varepsilon a_i},\ \ r_1 = \frac{h}{k}.$$

To determine $v_{i}j$, we have the equation

$$v_{i,j+1} = q^2_{ij}v_{ij} - p^2_{ij}v_{i+1,j} + \frac{f_{ij}}{t_j + \varepsilon} + O(k+h) \tag{36}$$
$$q^2_{ij} = 1 + \frac{\varepsilon a_i r - b_{ij}k}{\varepsilon + t_j},\ \ p^2_{ij} = \frac{\varepsilon a_i r}{t_j + \varepsilon}$$

From (34), we have the estimate

$$\left|c_{i,j+1}^{l}\right| \le \left|1 - \frac{\left[b_{ij} - b_i\right]k}{t_j + \varepsilon}\right|\left|c_{ij}^{l}\right| + O(k+h),$$

for sufficiently small $\varepsilon > 0$, for (35) we obtain

$$\left|c_{i+1,j}^{l}\right| \le \left|1 - \frac{\left[b_{ij} - b_i\right]h}{\varepsilon a_i}\right|\left|c_{ij}^{l}\right| + O(k+h). \tag{37}$$

From (36)

$$\left|v_{i,j+1}\right| \le \left|1 - \frac{b_{ij}k}{t_j + \varepsilon}\right|\left|v_{ij}\right| + \frac{f_{ij}}{t_j + \varepsilon} + O(k+h). \tag{38}$$

These estimates, according to the assumptions made at the beginning, imply the stability and convergence of schemes (34), (35), and (36) at a rate of $O(k+h)$. Solving problems (34)–(36), we find $c_{ij}{}^{l}, v_{ij}, i,j = \overline{1,n}$, using them we make a narrowing in (28) setting $x_i = (i-1)h,\ \ t_j = (j-1)k,\ \ i,j = \overline{1,n}$:

$$\tau = \tau_{i,j} = \varphi_{1,i}^{j} = \int_0^{t_j} \frac{1}{s+\varepsilon} b\left(A^{-1}\left(\varepsilon\left(z_{ij} + \ln\,(s+\varepsilon)\right)\right), 0\right) ds$$

$$\xi = \xi_{ij} = \varphi_{2,i}^{j} = \frac{1}{\varepsilon}\int_0^{x_i} \frac{b(s,0)}{a(s,0)} ds$$

define the solution of the original problem

$$u_{uv} = u_{ij} = v_{ij} + c_{i,j}^{1} e^{-\tau_{ij}} + c_{ij}^{2} e^{-\xi_{ij}} + \Phi\left(\xi_{ij} - \tau_{ij}\right) c_{ij}^{3} e^{-\tau_{ij}} + \Phi\left(\tau_{ij} - \xi_{ij}\right) c_{ij}^{4} e^{-\xi_{ij}}.$$

Following the methodology of works [38, 39] and based on estimates (37), (38), we obtain the estimate

$$\left|u(x,t,\varepsilon) - u_{uv}\right| < c(k+h).$$

Theorem 1.2 Let the given functions satisfy the above conditions. Then the solution constructed by the methods described above converges ε, uniformly at a rate of $O(k+h)$.

4. Conclusion

To construct an asymptotic solution of the problem posed with respect to regularizing functions, a first-order partial differential equation is obtained, the solution of which is described along the characteristic. One regularization variable is used to describe the boundary layer of power law ($\xi_1 = \varphi_1(x,t,\varepsilon)$), and the second regularization variable allows describing the exponential boundary layer along the straight line $x = 0(\xi_2 = \varphi_2(x,t,\varepsilon))$. To describe the corner boundary layer, an additional function $\Phi(\xi)$ is introduced, which allows describing the named boundary layer.

Acknowledgements

We thank the referees for taking the time to review and for their contributions to improve the clarity and impact of this paper.

Conflict of interest

The authors declare no conflict of interest.

Author details

Asan Sydygalievich Omuraliev[1,2†], Peyil Esengul Kyzy[1,2*†] and Ella Dayrbekovna Abylaeva[1,2]

1 Kyrgyz-Turkish Manas University, Bishkek, Kyrgyzstan

2 Kyrgyz State Technical University named after I. Razzakov, Bishkek, Kyrgyzstan

*Address all correspondence to: peyil.esengul@manas.edu.kg

† These authors contributed equally.

References

[1] O'Malley RE, Jr. Introduction to Singular Perturbations. 1st ed. Academic Press; 11 Feb 1974

[2] Babuska L, Osbom J. Analyls of finite element methods for second order boundary value problems using mesh dependent norms. Numerische Mathematik. 1980;**34**:41-62

[3] Babuska I, Szymazak WG. An error analysis for the finite element method applied to convection diffusion problems. Computation Methods Applied Mechanics and Engineering. 1982;**1**:19-42

[4] BAIL 1V. Proceedings the Fourth International Conference n Boundary and Interior Layers-Computational and Asymptotic Methods. Novosibirsk, USSR: Boole-Press; 1986

[5] Bank Burgler JF, Fichtner W, Smith RK. Some upwinding techlques for finite element approximations n convection-diffusion equations. Numerische Mathematik. 1990;**58**:185-202

[6] Barrett JW, Morton KW. Optimal finite element solutions to diffusionconvection problems In one dimension. International Journal for Numerical Methods in Engineering. 1980;**15**:1457-1474

[7] Bar-Yoseph, Israeli M. Asymptotic finite element method for improvement solutions boundary problems. Numerische Mathematik. 1986;**49**(4):425-438

[8] Berger. A conservative uniformly accurate difference method singular perturbation problem — Conservation form. IMA Journal of Numerical Analysis. 1986;**23**(6):1241-1253

[9] Berger AE. Solomon 1.M., Clment M. An analysis uniformly accurate difference method for singular perturbation problem. Mathematics of Computation. 1981;**37**(155):79-94

[10] Chen GZ, Yang ZF. A perturbatlonal h exponential unite difference scheme for the convectlve diffusion equation. Journal of Computational Physics. 1993; **104**:129-139

[11] de Groen PPN, Hemker PW. Error boimds for exponentlally fitted galerkin methods appld stiff two-point boundary value problems. In: Hemker PW, Miller JJH, editors. Numerical Analysis Singular Perturbation Problems. London: Academic Press; 1979. pp. 217-249

[12] Dorr PW, Parter SV, Shamplne LP. Applications the maximum principle to singular perturbation problems. SIAM Review. 1973;**I5**(1):43-88

[13] El-Mistikawy M, Werle MJ. Numerical method for boundary layers with blowing — The exponential box scheme. AIAA Journal. 1978;**16**:749-751

[14] Farrell P, O'Riordan JJ, Shishkin L. A uniformly convergent finite difference scheme for singularly perturbed semilinear equation. IMA Journal of Numerical Analysis. 1996;**33**(3):1135-1149

[15] Flaherty JE, Mathon W. In: Miller JJH, editor. Collocation Methods for Singularly Perturbed, Boundary Value Problems- //Boundary and Interior Lauers Computational and Asymptotic Methods. Dublin: Boole Press; 1980. pp. 77-92

[16] Gartland EG Jr. An analysis of a uniformly convergent finite difference - finite element scheme for model singular-perturbation problem.

Mathematics of Computation. 1988;**51**(183):93-106

[17] Kellogg RB. Analysis difference approximation for singular perturbation problem in two dimensions. In: Miller JJH, editor. Boundary and Interior Layerscomputational and Asymptotic Methods. Dublin: Boole Press; 1980. pp. 113-117

[18] Lin P-C, Guang-fu S. A completely exponentlally fitted difference scheme for a singular perturbation problem. Journal of Computational Mathematics. 1990;**8**(I):1-15

[19] Miller JJH, Shishkin. On the construction uniformly convergent finite difference schemes for singularly perturbed problems for quasilinear equation. In: Miller JJH, editor. Computational Methods for Boundary and Interior Layers Several Dimensions. Dublin: Boole Press; 1991. pp. 103-118

[20] Niijima K. On a difference scheme of exponential type for a nonlinear singular perturbation problem. Numerische Mathematik. 1985;**46**(4):521-539

[21] O'Rlordan. Singularly perturbed finite element methods. Numerische Mathematik. 1984;**44**(3):425-434

[22] Zhang Z. Finite element superconvergence approximation for onedimensional singularly perturbed problems. Numerical Methods for Partial Differential Equations. 2002;**18**:374-395

[23] Bakhvalov NS. Optimizatsiya metodov resheniya krayevykh zadach pri nalichii pogranichnogo sloya. ZHVM i MF. 1969;**9**(4):842-859

[24] Vasil'yeva AB, Butuzov VF. Asimptoticheskiye Metody v Teorii Singulyarnykh Vozmushcheniy. M.: Vysshaya shkola; 1990. 208 p

[25] Godunov SK, Ryaben'kiy VS. Raznostnyye skhemy. M: Nauka; 1977. p. 440

[26] Dulan E, Miller D, Shilders U. Ravnomernyye schetnyye metody resheniya zadach s pogranichnym sloyem. M: Mir; 1983

[27] Vasilyeva AB. On the internal transition layer in solving a system of the first order partial differential equations. Differential Equations. 1985;**21**:1537-1544. (in Russian)

[28] Nesterov AV, Shuliko OV. Asymptotics of the solution of a singularly perturbed system of partial differential equations of the first order with a small nonlinearity in the critical case. Journal of Computational Mathematics and Mathematical Physics. 2007;**47**(3):438-444. (in Russian)

[29] Nesterov AV. Asymptotics of the solution of the Cauchy problem for a singularly perturbed system of hyperbolic equations. Collection of Chebyshev. 2003;**12**(3):93-105. (in Russian)

[30] Nesterov AV. On the asymptotics of the solution of a singularly perturbed system of partial differential equations of the first order with a small nonlinearity in the critical case. Journal of Computational Mathematics and Mathematical Physics. 2012;**52**(7):1267-1276. (in Russian)

[31] Ilyin AM. Difference scheme for a differential equation with a small parameter at the highest derivative. Matematicheskie Zametki. 1969;**6**(2):237-248. (in Russian)

[32] Doolan ER, Miller JJH, Schilders WHA. Uniform Numerical Methods for Problems with Initial and Boundary Layers. Dublin: Boole Press; 1980

[33] Shishkin GI. Grid Approximations of Singularly Perturbed Elliptic and Parabolic Equations. Yekaterinburg: Publishing House of the Ural Branch of the Russian Academy of Sciences; 1992. (in Russian)

[34] Kellog RB, Tsan A. Analysis of some difference approximations for a singular perturbation problem without turning points. Mathematics of Computation. 1978;**32**(144):1025-1039

[35] Miller JJH, O'Riordan E, Shishkin GI. Fitted Numerical Methods for Singular Perturbation Problems. Singapore: World Scientific; 1996

[36] Bakhvalov NS. On optimization of methods for solving boundary value problems in the presence of a boundary layer. Zhurnal Vychislitel'noi Matematiki i Matematicheskoi Fiziki. 1969;**9**(4):841-859. (in Russian)

[37] Zadorin AI, Tikhovskaya SV. Analysis of a difference scheme for a singularly perturbed Cauchy problem on a refining grid. Sib. Magazine Comput. Mat. 2011;**14**(1):47-57. (in Russian)

[38] Shishkin GI, Shishkina LP. Improved difference scheme of the solution decomposition method for a singularly perturbed reaction-diffusion equation. Institute Mathematician and Mechanics of the Ural Branch of the Russian Academy of Sciences. 2010;**16**(1):255-271. (in Russian)

[39] Shishkin GI. Conditioning of a difference scheme of the solution decomposition method for a singularly perturbed convection-diffusion equation. Instituta Matematiki i Mekhaniki UrO RAN. 2012;**18**(2): 291-304

[40] Mishra HK, Saini S. Various numerical methods for singularly perturbed boundary value problems. American Journal of Applied Mathematics and Statistics. 2014;**2**(3): 129-142

[41] Lomov SA. Introduction to the General Theory of Singular Perturbations. Moscow: Nauk; 1981. (in Russian)

[42] Omuraliev AS. Numerical solution of a singularly perturbed initial problem method of a small parameter. In: Abstracts of Reports of the All-Union Conference. Nalchik; 1987. (in Russian)

[43] Omuraliev AS. Numerical regularization of a boundary value problem with a boundary layer arising at one end, bulletin of Osh state university. Ser. Physical Mathematics. 2001;**4**. (in Russian)

[44] Omuraliev AS. Numerical regularization of a singularly perturbed boundary value problem, Kyrgyz-Turk Manas University. MJEN. 2002;**2**:134-143. (in Russian)

[45] Omuraliev AS. Regularization of a singularly perturbed boundary value problem for an ordinary dif. equations based on finite elements. MJEN. 2003;**4**: 45-50. (in Russian)

[46] Omuraliev AS. On one finite element approach to solving a singularly perturbed problem, abstracts of the intern. In: Conference. According to Calc. Mat. ICVM-2004 June 21–25, Akademgorodok, Novosibirsk, Russia. 2004. (in Russian)

[47] Omuraliev AS. Numerical regularization of Cauchy problem for singularly perturbed parabolic equation, Kyrgyz-Turk Manas University. MJEN. 2004;**1**(5):1-5. (in Russian)

[48] Lomov SA. Power-law boundary layer in problems with singular

perturbation. Izvestiya Akademii Nauk SSSR. Seriya Matematicheskaya. 1966;**30**(3):525-572. (in Russian)

[49] Omuraliev AS, Esengul Kyzy P. Regularization of a singularly perturbed parabolic equation with power boundery layer, V congress of the Turkic world mathematicians, Kyrgyzstan. Issyk-Kul Aurora. 2014;**5-7**(June):136-142

[50] Omuraliev AS, Abylaeva ED, Esengul Kyzy P. A system of singularly perturbed parabolic equations with a power boundary layer. Lobachevskii Journal of Mathematics. 2020;**41**(1):71-79

[51] Omuraliev AS, Abylaeva ED, Esengul Kyzy P. A parabolic problem with a power-law boundary layer. Differential Equation. 2021;**57**(1):67-77

[52] Ladyzhenskaya OA, Solonnikov VA, Uraltseva NN. Linear and Quasilinear Equations of Parabolic Type. Moscow: Nauka; 1967. (in Russian)

Chapter 7

Fundamental Differential Equations in Mathematical-Physics Using the Operator Method with Applications in Laser Beam and Electron Beam Processing

Claudiu Hapenciuc, Andrei Popescu, Cristian N. Mihăilescu and Mihai Oane

Abstract

In the last 25 years, a semi-analytical-numerical method has been developed to solve the heat equation using integral transform technique. The solution of differential equations was applied successfully to a large area of laser processing phenomena. The model was applied successfully to a large area of phenomena in laser and electron beam processing. After this stage, we have used another method to solve the same phenomena. We are using the operator method which is an analytical model, developed mainly by K. Zhukovsky. In the present chapter, we will give an example on how to apply the operator method on heat phenomena during laser-metal interaction. In the second part of our chapter, we will also deal with thermal field in starch irradiated by electron beam using operator method.

Keywords: partial differential equations, operator method, non-Fourier equations, laser-metal interaction, electron beam-starch interaction

1. Introduction

We have successfully used the integral transform technique with direct applications in laser processing. We can mention few important contributions, like references [1–12]. The integral transform technique is part of the great family of mathematical methods of eigenfunctions and eigenvalues. It is characterized by the fact that it can take into account a large spectrum of spatial boundary conditions and also almost any kind of laser-target interaction. It is the "place" to mention that the solutions of the above-mentioned method is of the form of the sum of millions of functions, which is in fact the sum of the terms of a polynomial with a huge number of elements. We want to mention the fact that the solutions of the form of polynomials are very well "fitted" with the use of the software MATHEMATICA. On the other hand, the differential operator method which was developed in the last decade mainly by K. Zhukovsky [13–20] is based on solving partial differential equations using operators like in standard non-relativistic quantum mechanics, the most important being the heat operator. The boundary conditions are based on the initial time conditions. The method is very powerful for the situations where ultra-fast interactions are involved: 10^{-21} s to 10^{-9} s.

Polynomials are omnipresent in today's theoretical physics. For example, the final solutions (mainly temperatures) using integral transform technique and/or differential operator are of the form of polynomials. For example, using the integral transform technique, we have an infinite sum of polynomials, and therefore, we are in the situation to take just first 100–1.000.000 terms because hopefully the solutions are convergent. In consequence, the solution becomes from 100% analytical to a semi-analytical-numerical one. In the case of differential operator method, the situation is easier, the solution being 100% analytical. In this situation, we have only polynomial results. We may say that apart from 100% numerical methods, the other theoretical methods make use of polynomials. During ultrashort laser pulse interaction with metals, the heat transfer proceeds in two steps. First, energy absorption occurs via photon-electron interactions, followed by return of excited electrons to the initial state, during a few femtoseconds. Next, the energy is redistributed from electrons to lattice by electron-phonon interactions, in a frame of few picoseconds, followed by thermalization, i.e., the heat is dissipated, and the lattice reaches thermal equilibrium [21–23]. For the case of nanoparticle interaction with ultra-short laser pulse, the problem is more delicate, especially when we consider like in our case attosecond pulses, i.e., the laser pulse duration is shorter than the relaxation time. For this goal, one may appeal to the notions of quantum field theory.

2. The Kozłowski thermal model

In the present paragraph, we will briefly describe the Kozłowski thermal model, proposed in Refs. [24, 25]. The research of transport mechanisms at the nanoscale is of huge importance nowadays. In particular, the nanoparticles and nanotubes have important physical applications for nanoscale heat transfer. A lot of models have been developed in the simple picture of point-like particles. The development of the ultra-short electromagnetic pulses opens new possibilities in the study of the dynamics of the electrons in nanoscale systems: carbon nanotubes and nanoparticles. For attosecond laser pulses, the duration of the pulse is shorter than the relaxation time. In that case, the transport equations contain the second-order partial derivative in time. In this paragraph, we consider the nonlinear Klein–Gordon equation for mass and thermal energy transport in nanoscale. Considering the results of the monograph [24], we develop the nonlinear Klein–Gordon equation for heat and mass transport in nanoscale. In Ref. [25], it was shown that in the case of the ultra-short laser pulses, the heat transport is described by the hyperbolic heat transport equation:

$$\tau \frac{\partial^2 T}{\partial t^2} + \frac{\partial T}{\partial t} = D\left(\frac{\partial^2 T}{\partial x^2} + \frac{\partial^2 T}{\partial y^2} + \frac{\partial^2 T}{\partial z^2}\right) \tag{1}$$

where T denotes the temperature of the electron gas in nanoparticle, τ is the atomic relaxation time, and D is the thermal diffusion coefficient. The relaxation time is defined as follows:

$$\tau = \frac{\hbar}{mv^2}, v \equiv \alpha c \tag{2}$$

where v is the thermal pulse propagation speed. For electromagnetic interaction when scatters are the relativistic electrons, v = c, Thomson relaxation time, m is the electron mass:

$$\tau = \frac{\hbar}{mc^2}, \tag{3}$$

Both parameters τ and v completely characterize the thermal energy transport on the atomic scale and can be named as "atomic" relaxation time and "atomic" heat velocity.

In the following, starting with the atomic τ and v, we describe thermal relaxation processes in nanoparticles which consist of N light scatter points.

The number (N) of particles in an Au nanoparticle (sphere with radius {XE "radius"}r) can be calculated using the formula:

$$N = \frac{\frac{4\pi r^3}{3}}{\mu} \rho A Z \propto r^3 \tag{4}$$

where ρ is the density of the nanoparticle, A is the Avogadro number, μ is the molecular mass {XE "molecular mass"} of the particles in grams, and Z is the number of {XE "valence"} electrons {XE "electrons"}. Here, r denotes the radius of the nanoparticle, and p is the momentum of energy carriers.

According to Kozłowski, we recalculate the relaxation time τ for nanoparticles consisting N electrons:

$$\tau^N \propto r^3 \tau \tag{5}$$

With formulae (4) and (5), the heat {XE "transport"} equation takes the form:

$$r^3 \tau \frac{\partial^2 T}{\partial t^2} + \frac{\partial T}{\partial t} = \frac{\hbar^{\frac{1}{3}}}{m} \left(\frac{\partial^2 T}{\partial x^2} + \frac{\partial^2 T}{\partial y^2} \frac{\partial^2 T}{\partial z^2} \right) \tag{6}$$

Eq. (6) is linear damped Klein–Gordon equation and was solved for nanotechnology {XE "nanotechnology"} systems in [24, 25] using the nonlinear d'Alembert equation. We will use another path, using Zhukovsky mathematical models.

3. The Zhukovsky mathematical model

The thermal Klein–Gordon equation in Zhukovsky model contains the heat equation and thermal wave equation in explicit form:

$$\left(\tau \partial_t^2 + \partial_t\right) T(x, y, z, t) = \left(D \bullet \nabla^2 + k\right) T(x, y, z, t) \tag{7}$$

Here, D represents the Au thermal diffusivity; $D = \frac{K}{C_P \rho} = \frac{\alpha}{} = \alpha\tau$, τ is the thermal relaxation time, C_P is heat capacity, K is thermal conductivity, and ρ is mass density of the irradiated sample. k is a real number (not to be confused with thermal conductivity). According to reference [26], we have the following solution:

$$T(x,y,z,t) = e^{-\frac{\varepsilon t}{2}} \cdot \frac{t}{4\sqrt{\pi}} \int_0^{\infty} \frac{d\xi}{\xi\sqrt{\xi}} \cdot e^{-\frac{t^2}{16\,\xi} - \xi^2\left(\varepsilon^2 + 4k\right)} \hat{S}_x \hat{S}_y \hat{S}_Z f(x,y,z) \tag{8}$$

with

$$f(x,y,z) = T\ (x,y,z,0) \tag{9}$$

the initial temperature, which in Zhukovsky model, plays the role of source term. In consequence, the shorter the irradiation is, the better approach is the Zhukovsky model.

We also define the heat operators:

$$\hat{S}_i = e^{-4\alpha\xi\partial_i^2} f(i); .i = x,y,z \tag{10}$$

For simplicity, we will set $k = 0$ and $f(x,y,z) \equiv \delta(x,y,z)$.

And we have the final temperature field expression [26]:

$$T(x,y,z,t) = e^{-\frac{\varepsilon t}{2}} \cdot \frac{t}{64\pi^2\alpha^{\frac{3}{2}}} \int_0^{\infty} \frac{d\xi}{\xi^3} \cdot e^{-\xi\epsilon^2 - \frac{t^2 - \left(x^2+y^2+z^2\right)/\alpha}{16\xi}} \tag{11}$$

4. The link between Kozłowski thermal model and Zhukovsky mathematical model

In order to make a link between Kozłowski thermal model and Zhukovsky mathematical models, we have to make an analogy between Eqs. (1) and (7).

If we set $k = 0$ in Eq. (7), we have the following "correspondences":

$$\tau \rightarrow r^3\tau \tag{12}$$

And

$$D \rightarrow \frac{\hbar^{\frac{1}{3}}}{\mathrm{m}} \tag{13}$$

In our case, it is reasonable, due to the fact that we have ultra-short laser pulses (which will affect mostly the nanoparticle surface), and on the other hand, the fact that the target is small (radius 100 nm and 50 nm) to take the following expression for Eq. (9):

$$f(x,y,z) = T\ (x,y,z,0) \propto r^2\delta(x,y,z) \tag{14}$$

In this case, we have the following formula for electron's temperature of a nanoparticle under ultra-short laser pulse irradiation (from Eqs. (11)–(14)):

$$T(x,y,z,t) = e^{-\frac{t}{2r^3\tau}} \cdot \frac{r^2 t}{64\pi^2\left(\frac{\hbar^{\frac{1}{3}}}{mr^3\tau}\right)^{\frac{3}{2}}} \int_0^{\infty} \frac{d\xi}{\xi^3} \cdot e^{-\xi/r^6\tau^2 - \frac{t^2 - \left(x^2+y^2+z^2\right)/\left(\frac{\hbar^{\frac{1}{3}}}{mr^3\tau}\right)}{16\xi}} \tag{15}$$

5. The simulations based on the thermal Klein: Gordon equation

In our simulations, we will examine an Au nanoparticle under attosecond laser pulse irradiation. We will take the "classical" Au relaxation time equal to 10^{-13} s. For the radius of a nanoparticle, we will consider a value of 50 nm (corresponding to a diameter of 100 nm). The surrounding medium for the Au nanoparticle will be water. In our simulations, the only parameter which was taken equal with 1 a.u. was the laser intensity. The irradiation time was considered 100 as in the whole our study. In our simulations, we considered that y = z = 0, and x is the direction of laser propagation.

Figures 1 and **2** represent the electron thermal fields versus time and x coordinate in the first fs after an irradiation of 100 as of a nanoparticle with radius of 100 nm, respectively, first 10 fs. **Figure 3** presents electron thermal field versus time and x coordinate in the first fs after an irradiation of 100 as of a nanoparticle with a radius of 50 nm. Finally **Figure 4** presents the thermal field coordinate in the first fs after an irradiation of 100 as of a nanoparticle with a radius of 100 nm during the first 100 as. All the arbitrary units in **Figures 1–4** are self-consistent; in consequence, we can compare the graphics between them. We observe in **Figure 1** that after 1 fs, the temperature is high and mainly at the surface of the nanoparticle. In **Figure 2**, it can be observed that the water medium absorbed all the nanoparticle's heat. In consequence, the temperature, as shown in **Figure 2**, goes rapidly to zero. We have to mention that, in **Figures 1–4** all the temperatures represent temperature variation rather than absolute temperature. In **Figure 3**, we present the same field similar to in **Figure 1** with the only difference that in **Figure 3**, we have a radius of 50 nm. The electron thermal field versus time and x coordinate in the first 100 as after an irradiation of 100 as of a nanoparticle with radius of 100 nm is presented in **Figure 4**. We observe that we still have a substantial and detectable temperature field. In fact, during 100 as irradiation, the temperature is very large in comparison with **Figure 1**. Also, the temperature decreases drastically from about 25,000 a.u. (after 100 as) to 25 a.u. (after 1 fs) and to almost 0 (after 10 fs). We may comment also that when the radius is reduced to 50 nm, the temperature decreases approximatively 10 times (comparison between **Figures 1** and **3**). This is due to the fact that the radius plays an

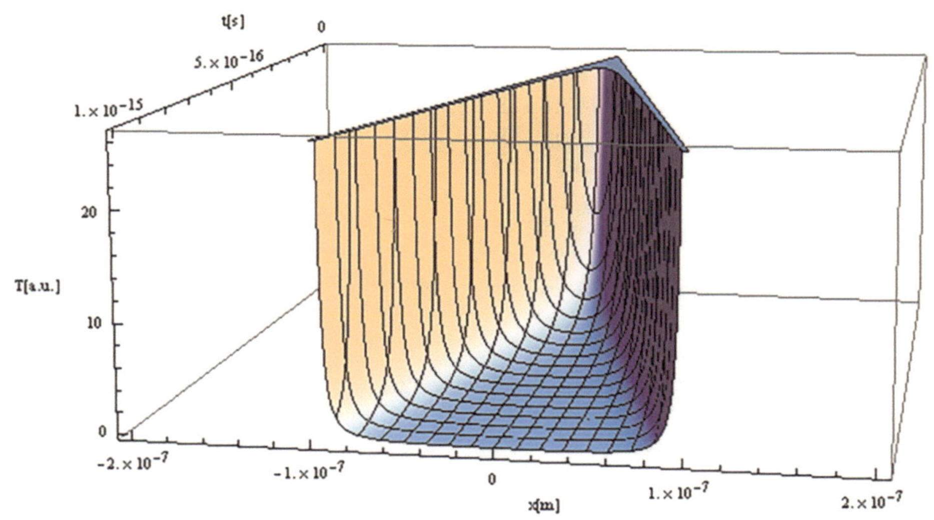

Figure 1.
Electron thermal field versus time and x coordinate (y = z = 0) in the first fs after an irradiation of 100 as of a nanoparticle with radius of 100 nm.

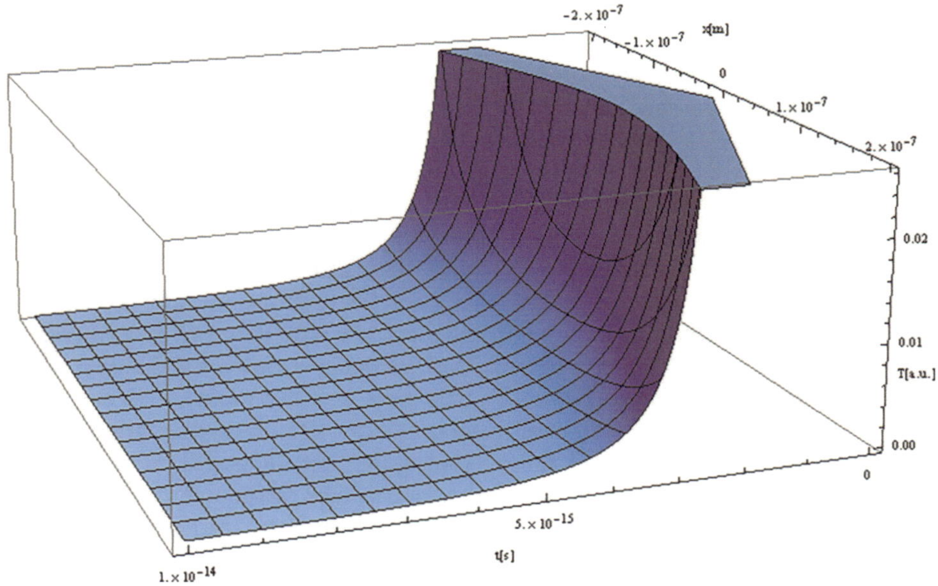

Figure 2.
Electron thermal field versus time and x coordinate (y = z = 0) in the first 10 fs after an irradiation of 100 as of a nanoparticle with radius of 100 nm.

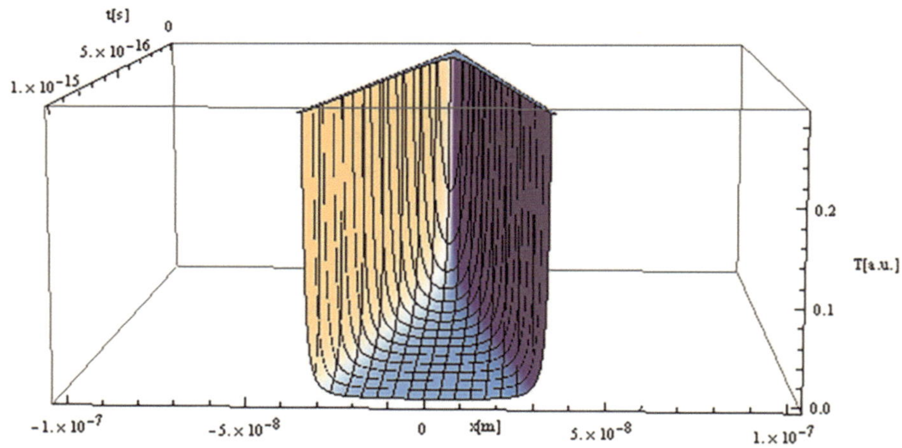

Figure 3.
Electron thermal field versus time and x coordinate (y = z = 0) in the first fs after an irradiation of 100 as of a nanoparticle with radius of 50 nm.

important role in the temperature: the higher the radius is, the higher is the temperature variation.

There are very few ways to solve the heat equation. For example, the integral transform technique was applied with success long time ago [9, 10], but in our attempt to apply the mentioned formalism to ultra-short scale (distance or time), the results were not very impressive.

On the other way, the Kozlovsky formalism is excellent for nanoparticles and for ultra-short in time pulses. We have applied in the present paper successfully the

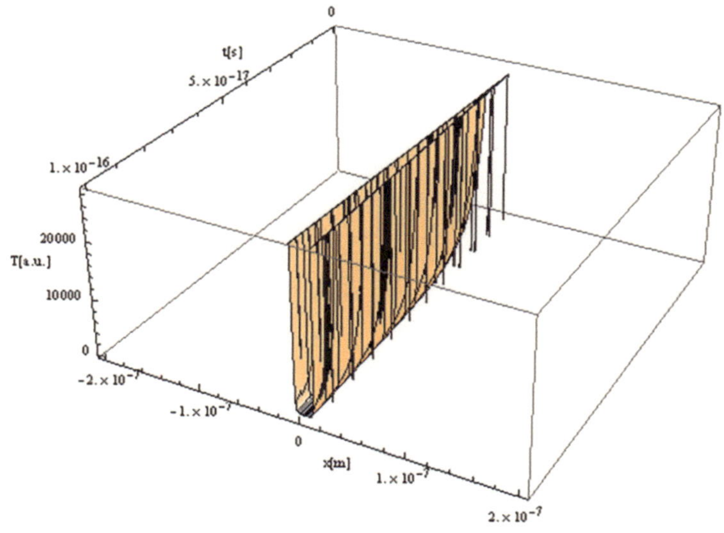

Figure 4.
Electron thermal field versus time and x coordinate (y = z = 0) in the first 100 as after an irradiation of 100 as of a nanoparticle with radius of 100 nm.

Kozlovsky technique for flash laser experiments on nanoparticles, giving a rigorous solution to 3D thermal Klein–Gordon heat equation [24, 25].

6. Experiment versus theoretical computer modeling: Electron beam irradiation of starch

In Refs. [13–20], the operational method was used in order to construct particular solutions for the heat equation. The formalism uses inverse differential operators and also operational exponentials. Zhukovsky formalism [13–20] provides a very strong solution applicable to real situations. Heat conduction is an important transport process which is governed by conservation of energy or in other words the first law of thermodynamics. By applying the Zhukovsky mathematical model, the first law of thermodynamics leads to the linear generalized Fourier model [14], namely Fourier–Zhukovsky heat transfer model. We can apply the Zhukovsky formalism, also to non-Fourier (telegraph equations) extensions [15]. This is the base of our simulations, which we will compare with thermal field from experimental situation during relativistic electron beam irradiation of maize starch. The thermal properties of different kind of starch have been investigated by various research groups some time ago [26–28]. In the present paper, we evaluate the thermal field of maize starch under the irradiation with the ALID 5.5 MeV electron beam facility from INFLPR, Romania. It is essential to say that the temperature appearing in all figures is in fact the absolute temperature expressed in Celsius degrees. In order to have a best fit between experimental data and theoretical simulations, we have chosen for maize starch the following parameters: c = 3.06KJ/ (Kg K), k = 0.078 W/ (m K), and h = 0.02 W/ (mm^2 K) which express heat capacity, heat conductivity and, respectively, heat transfer coefficient. The values of heat capacity and heat conductivity are of the order of magnitude with those presented in literature [26–29]. We have done irradiations on a plastic sample, and we have measured the temperature field in order to find the shape of the relativistic electron beam which is a key factor in determining the source term (**Figures** 5 and **6**). The experimental data are presented in **Figures** 7 and **8**.

The function which fits with a good approximation the electron beam intensity can be written as:

$$I = h + j \times x^3 + c \times x^2 + d \times x + e + k \times y^3 + f \times y^2 + g \times y \tag{16}$$

where j = $1.6258788130963823\text{X}10^{-6}$.
c = −0.000147127208979571.
d = −0.005256265596577318.

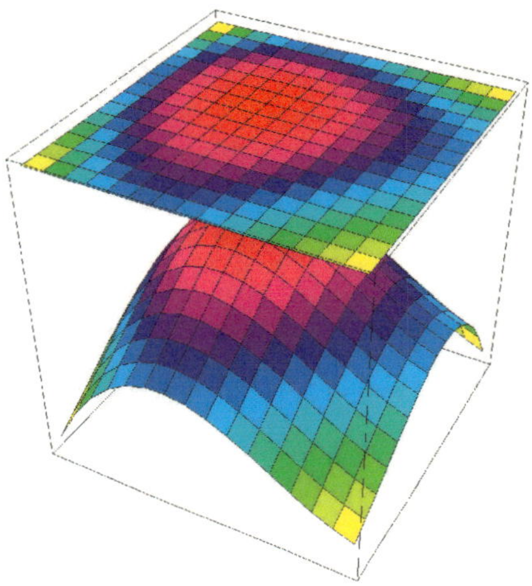

Figure 5.
3D and projection of the thermal field during electron beam irradiation of the plastic target.

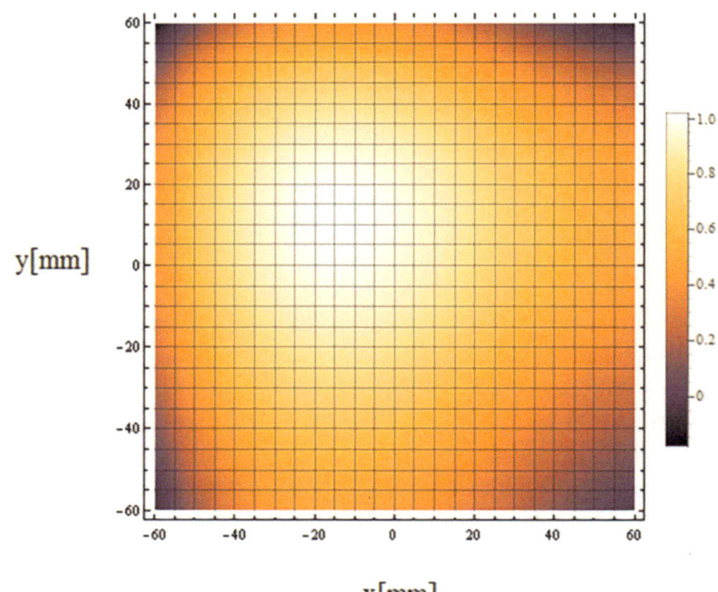

Figure 6.
2D normalized thermal field. The shape of the electron beam intensity is close to a Gaussian one.

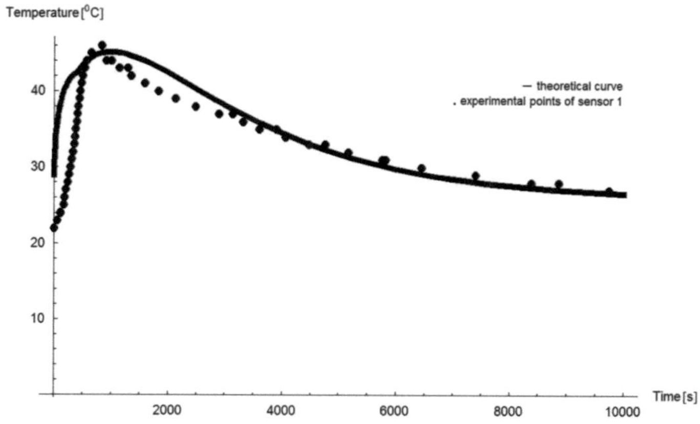

Figure 7.
Experimental data (dots) versus computer simulations (line) for the sensor 1.

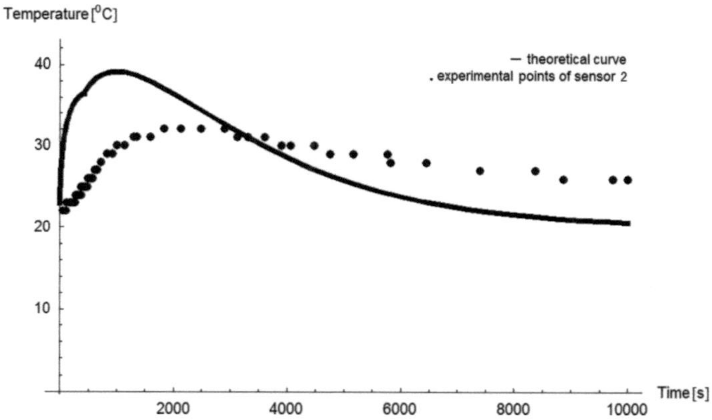

Figure 8.
Experimental data (dots) versus computer simulations (line) for the sensor 2.

e = 0.402069778147499 17.
k = $-1.2507174544855288\text{X}10^{-6}$.
f = −0.00014753303521 77021.
g = 0.0038280551388935637.
h = 0.61927959821428556.

We have $f\left(x, y\right) \propto I$. From the experimental point of view, we observe that we have three thermal regimes: (i) temperature increase during irradiation (435 s), (ii) increase of temperature during post-irradiation regime (435 s–833 s), and iii) cooling process (920 s − 10,000 s).

Starch from corn (S4126; moisture content: ≤15%) was purchased from Sigma-Aldrich Company.

The starch powder was placed in a cylindrical cardboard box, which can be described in Cartesian coordinates.

The temperatures were measured at the back side of the maize-starch target, in the center of the target x = y = 0, z = 55 mm, for the sensor 1 and x = 0, y = 35 mm, and z = 55 mm for the sensor 2.

We have

$$f(x,y,z,t) = D'(t)\rho\big(f_1(x) + f_2(y)\big) \cdot f_3(z) \tag{17}$$

Here, $D'(t)$ is doze debit of the incident electron beam, and ρ is the target mass density. For simplicity, we will include these both constants in normalized coefficients.

where

$$f_1(x) = h + j \cdot x^3 + c \cdot x^2 + d \cdot x \tag{18}$$

$$f_2(y) = \mathrm{e} + k \bullet y^3 + f \cdot y^2 + g \cdot y \tag{19}$$

Also, we may assume

$$f_3(z) = 1 - \frac{z}{R} \tag{20}$$

With $R = 5,5\,cm$

7. The mathematical formalism

The telegraph equation contains in explicit form the heat equation and thermal wave equation:

$$\left(\tau\partial_t^2 + \partial_t\right)T(x,y,z,t) = \left(D_T\nabla^2 + k\right)T(x,y,z,t) \tag{21}$$

Here, $D_T = 49,2\ \frac{mm^2}{s}$ represents the thermal diffusivity; $D_T = \frac{K}{C_P\rho} = \frac{\alpha}{} = \alpha\tau$, τ is the thermal relaxation time, C_P is heat capacity, K is thermal conductivity, and $\rho =$ 0,541 $\frac{g}{Kcm^3}$ is mass density of the irradiated sample. k is a real number (not to be confused with thermal conductivity) . According to reference [14], we have the following solution:

$$T(x,y,z,t) = \mathrm{e}^{-\frac{\varepsilon t}{2}} \cdot \frac{t}{4\sqrt{\pi}} \int_0^\infty \frac{d\xi}{\xi\sqrt{\xi}} \cdot \mathrm{e}^{-\frac{t^2}{16\,\xi} - \xi^2\left(\varepsilon^2 + 4k\right)} \hat{S}_x \hat{S}_y \hat{S}_Z f(x,y,z) \tag{22}$$

With

$$f(x,y,z) = T\,(x,y,z,0) \tag{23}$$

the initial temperature, which in Zhukovsky model plays the role of source term. In consequence, the shorter the irradiation is, the better approach is the Zhukovsky model. In our experiment, the irradiation time represents 4.3% from the total experiment time.

We have

$$\hat{S}_i = \mathrm{e}^{-4\alpha\xi\partial_i^2} f(i); i = x,y,z \tag{24}$$

According to Eqs. (17)–(20) and (24), we obtain

$$\hat{S}_x = e^{-4\alpha\xi\, \partial_x^2 (f_1(x)+f_2(y))f_3(z)} \tag{25}$$

$$\hat{S}_y = e^{-4\alpha\xi\, \partial_y^2 (f_1(x)+f_2(y))f_3(z)} \tag{26}$$

$$\hat{S}_z = e^{-4\alpha\xi\partial_z^2 (f_1(x)+f_2(y))f_3(z)} \tag{27}$$

In consequence, from Eq. (25) we obtain

$$\hat{S}_x = e^{-4\alpha\xi\partial_x f_3(z)\partial_x f_1(x)} \tag{28}$$

$$\hat{S}_x = e^{-4\alpha\xi f_3(z)\partial_x^2 f_1(x)} \tag{29}$$

And similarly,

$$\hat{S}_y = e^{-4\alpha\xi f_3(z)\partial_y^2 f_2(y)} \tag{30}$$

$$\hat{S}_z = e^{-4\alpha\xi(f_1(x)+f_2(y))\partial_z^2 f_3(z)} \tag{31}$$

From Eqs. (14)–(16), we derive

$$\hat{S}_{x,y,z} f(x,y,z) = \hat{S}_x \cdot \hat{S}_y \cdot \hat{S}_z f(x,y,z) = e^{-4\alpha\xi\,\left(1-\frac{z}{R}\right)\cdot(2C+6jx)\cdot} e^{-4\alpha\xi\left(1-\frac{z}{R}\right)(2f+6ky)} \tag{32}$$

The final solution, after an irradiation time t_0, is

$$\begin{aligned} T(x,y,z,t) &= e^{-}\frac{\mathcal{E}t}{2} \cdot \frac{t}{4\sqrt{\pi}} \int_0^\infty \frac{d\xi}{\xi\sqrt{\xi}} \cdot \\ & e^{-\frac{t^2}{16\,\xi} - \xi^2\left(\mathcal{E}^2+4k\right)} e^{-4\alpha\xi\,\left(1-\frac{z}{R}\right)\cdot(2C+6jx)\cdot} e^{-4\alpha\xi\left(1-\frac{z}{R}\right)(2f+6ky)} (h(t)-h(t-t_0)) \\ & + e^{-\frac{\mathcal{E}t}{2}} \cdot \frac{t}{4\sqrt{\pi}} \int_0^\infty \frac{d\xi}{\xi\sqrt{\xi}} \cdot e^{-\frac{t^2}{16\,\xi} - \xi^2\left(\mathcal{E}^2+4k\right)} h(t_0) \end{aligned} \tag{33}$$

In our experiment, we have irradiated maize starch with 5.5 MeV electron beam. The irradiation time was 435 s. In order to have a very good concordance between experiment and computer modeling, we have set $\tau = 10^3$s (the heat relaxation time). One can notice that, after irradiation time, we still have a temperature increase. This is due to the fact that our final solution takes into account also the thermal waves.

8. General conclusions

We may conclude that the differential operator method is a very powerful one, because it can be applied in a lot of different situations from free electrons laser to laser processing. The formalism is not heavy to be implemented in software form, in our case MATHEMATICA. The running time, on the other hand, is about less than 1 minute. The simulations results are in good agreement with the experimental data from literature. The method is a state of the art one, being developed in the last decade. In the end, we have to mention which method is more powerful and in which respects. For the very beginning, as we already mentioned, it is still an issue for the future to succeed in combining both models: the semi-analytical-numerical integral transform technique and analytical differential operator technique. The first model, which was developed in 1950, when the laser was not yet been discovered is very powerful with respect to the spatial boundary conditions, while the second one developed in 2010–2020 is very powerful in taking into account very short periods of time, which is helpful in ultra-short lasers and cosmology (to name just two areas of applications).

At the end of the present chapter, it is an opportunity to mention some specific and realistic future achievements of the differential operator technique. With the integral transform technique, we had success to take very complicated quantum interaction like, for example, multi-photon processes [10], but at reasonable periods of time, such as 10^{-8} s, but for 10^{-18} s, we need to use the differential operator technique. For what is it such accuracy necessary in experimental physics? The answer is very simple and clear to extend the two-temperature model [10] from laser–metal interaction to laser–dielectric and laser–semiconductor interactions. Also, one can solve using both methods presented in the present paper for determining the thermal field during electron beam-matter interaction [27–31].

Author details

Claudiu Hapenciuc, Andrei Popescu*, Cristian N. Mihăilescu and Mihai Oane
National Institute for Laser, Plasma and Radiation Physics (INFLPR), Măgurele-Ilfov, Romania

*Address all correspondence to: andrei.popescu@inflpr.ro

References

[1] Oane M, Sporea D. Temperature profiles modelling in IR optical components during high power laser irradiation. Infrared Physics and Technology. 2001;**42**(1):31-40

[2] Oane M, Apostol D. Mathematical modelling of two-photon thermal fields in laser-solid interaction. Optics and Laser Technology. 2004;**36**:219-222

[3] Oane M, Scarlat F, Mihailescu IN. Temperature field modeling during multi- modes CO2 laser irradiation of human enamel. Optics and Laser Technology. 2007;**39**:537-540

[4] Oane M, Tsao SL, Scarlat F. Temperature field distribution in multi - layered solid media heated with multi-modes laser beam. Optics and Laser Technology. 2007;**39**:179-181

[5] Oane M, Scarlat F, Tsao SL, Mihailescu IN. Thermal fields in laser-multi-layer structures interaction. Optics and Laser Technology. 2007;**39**: 796-799

[6] Oane M, Peled A, Scarlat F, Mihailescu IN, Scarisoreanu A, Georgescu G. Multi- photon temperature profile modeling in solids during powerful pulse laser irradiation. Infrared Physics and Technology. 2008;**51**: 242-245

[7] Oane M, Peled A, Scarlat F, Mihailescu IN, Georgescu G. Quantum Fourier models for semiconductors under multiple laser irradiations. Infrared Physics and Technology. 2008; **51**:348-350

[8] Oane M, Scarlat F, Mihailescu IN. The semi-analytical solution of the Fourier heat equation in beam- 3D inhomogeneous media interaction. Infrared Physics and Technology. 2008; **51**:344-347

[9] Oane M. Laser calorimetry in action: Classical versus quantum physics? Lasers in Engineering. 2010;**20**(5–6):329-345

[10] Oane M, Taca M, Tsao SL. Two temperature models for metals: A new "radical" approach. Lasers in Engineering. 2012;**24**(1–2):105-113

[11] Braşoveanu M, Oane M, Nemţanu MR. Heat transport in starch exposed to ionizing radiation: Experiment versus theoretical computer modeling. Starch/Starke. 2019;**71**(11-12): 1900147

[12] Mihăilescu CN, Oane M, Mihăilescu N, Ristoscu C, Mahmood MA, Mihăilescu IN. Matrix theory - classics and advances. In: A New Approach to Solve Non-Fourier Heat Equation Via Empirical Methods Combined with the Integral Transform Technique in Finite Domains. Croatia: IntechOpen; 2022. DOI: 10.5772/ intechopen.104499

[13] Zhukovsky K. Operational method of solution of linear non-integer ordinary and partial differential equations. Springer-Plus. 2016;**5**:119. DOI: 10.1186/ s40064-016-1734-3

[14] Zhukovsky K. Operational approach and solutions of hyperbolic heat conduction equations. Axioms. 2016;**5**:28

[15] Zhukovsky K, Srivastava HM. Analytical solutions for heat diffusion beyond Fourier law. Applied Mathematics and Computation. 2017; **293**:423-437

[16] Zhukovsky K. Violation of the maximum principle and negative

solutions with pulse propagation in Guyer–Krumhanslmodel. International Journal of Heat and Mass Transfer. 2016; **98**:523-529

[17] Zhukovsky K. Exact solution of Guyer–Krumhansl type heat equation by operational method. International Journal of Heat and Mass Transfer. 2016; **96**:132-144

[18] Zhukovsky K. Exact negative solutions for Guyer–Krumhansl type equation and the violation of the maximum principle. Entropy. 2017;**19**:440

[19] Zhukovsky K. A harmonic solution for the hyperbolic heat conduction equation and its relationship to the Guyer-Krumhansl equation. Moscow University Physics Bulletin. 2018;**73**(1): 45-52. DOI: 10.3103/S0027134918010186

[20] Zhukovsky K. Exact harmonic solution to ballistic type heat propagation in thin films and wires. International Journal of Heat and Mass Transfer. 2018;**120**:944-955

[21] Dyukin RV, Martsinovskiy GA, Sergaeva ON, Shandybina GD, Svirina VV, Yakovlev EV. In: Peshko I, editor. Interaction of Femtosecond Laser Pulses with Solids: Electron/Phonon/ Plasmon Dynamics, Laser Pulses - Theory, Technology, and Applications. Croatia: IntechOpen; 2012

[22] Shin T, Teitelbaum S, Wolfson J, Kandyla M, Nelson L. Extended two-temperature model for ultrafast thermal response of band gap materials upon impulsive optical excitation. The Journal of Chemical Physics. 2015;**143**:194705

[23] Jiang L, Tsai HL. Improved two-temperature model and its application in ultrashort laser heating of metal films. Journal of Heat Transfer. 2005;**127**: 1167-1173

[24] Kozłowski M, Marciak-Kozłowska J. Thermal Processes Using Attosecond Laser Pulses. Germany: Springer; 2006. ISBN: 0387301593

[25] Kozłowski M, Marciak-Kozłowska J. From Femto-To Attoscience and Beyond. United Kingdom: Nova Science Pub Inc; 2009. ISBN: 1607411644

[26] Oane M, Ticoş D, Ticoş CM Scholars' Press - Germany, Charged Particle Beam Processing Versus Laser Processing. 2015

[27] Tana I, Weeb CC, Sopadea PA, Halleya PJ. Estimating the specific heat capacity of starch water-glycerol systems as a function of temperature and compositions. Starch/Stärke. 2004;**56**: 6-12

[28] Drouzas AE, Saravacos GD. Effective thermal conductivity of granular starch materials. Journal of Food Science. 1988;**53**(6):1795-1799

[29] Crăciun G, Mănăilă E, Stelescu MD. New elastomeric materials based on natural rubber obtained by electron beam irradiation for food and pharmaceutical use. Materials. 2016; **9**(12):999

[30] Oane M, Toader D, Iacob N, Ticoş CM. Thermal phenomena induced in a small graphite sample during irradiation with a few MeV electron beam: Experiment versus theoretical simulations. Nuclear Instruments and Methods in Physics Research B. 2014; **318**:232-236

[31] Oane M, Toader D, Iacob N, Ticoş CM. Thermal phenomena induced in a small tungsten sample during irradiation with a few MeV electron beam: Experiment versus simulations. Nuclear Instruments and Methods in Physics Research B. 2014;**337**:17-20

Chapter 8

Implications of Higher-Degree Polynomials in Forced Damped Oscillations

Lucian Milica

Abstract

Exploring fundamental mathematics, especially polynomials, provides clear insights into the structure and behavior of different systems. Polynomials are algebraic expressions that enable the formulation and solution of a wide range of mathematical problems, from simple equations to complex motion models. An exciting example of the application of polynomials in physics and engineering is the analysis of mechanical vibrations in a forced damped oscillating system, such as those encountered in ball or roller bearings. Such a system is described by a set of linear differential equations governing the motion of its components. The mathematical model presented in this chapter describes a system with two degrees of freedom, whose movement is expressed by a system of two second-order differential equations. In solving these equations, a matrix approach was used to group the corresponding terms of the first and second derivatives of the displacements. By properly identifying these terms, a system of matrix equations was obtained that can be solved to determine the constant coefficients according to the particular solutions of the system, providing a deeper understanding of its dynamic behavior. The proposed mathematical model is a general one that can be applied to any similar oscillatory system.

Keywords: damped forced vibrations, characteristic polynomial, differential equations, natural frequencies, vibrations of radial bearings

1. Introduction

Polynomials play a crucial role in mathematics and science, and their exploration provides an essential framework for understanding and applying mathematics in various branches of industry. From solving equations to modeling natural phenomena and approximating functions, polynomials are a fundamental element of the mathematical language, providing a simple and efficient way to express and analyze a wide range of mathematical and scientific problems. One of these problems is represented by the vibrations produced in the bearings.

The level of vibrations in bearings is considered an important quality index with numerous implications. On the one hand, these vibrations directly affect the performance and competitiveness of equipment within an application; on the other hand, they underscore the dimensional and form precision in manufacturing, as well as the efficiency of finishing, assembly, and maintenance technologies used in bearing production. In the case of high-precision devices, an important factor in the occurrence of vibrations is the uneven rotation of the bearing, which leads to negative consequences in their kinematic chain.

Radial ball bearings are a type of rolling-element bearing designed to support loads perpendicular to the axis of rotation, known as radial loads.

Here are the key components of radial ball bearings (**Figure 1**).

1. Inner ring;

2. Outer ring;

3. Rolling elements;

4. Cage (roller retainer).

The inner ring provides the raceway for the rolling elements (balls or rollers). It is usually mounted on the rotating shaft. The outer ring provides the stationary raceway for the rolling elements. It is typically housed in a bearing housing or support structure. Rolling elements can be either balls or rollers. They roll within the raceways of the inner and outer rings, distributing the load and reducing friction. The choice between balls and rollers depends on the specific application requirements. The cage separates and guides the rolling elements, preventing them from contacting each

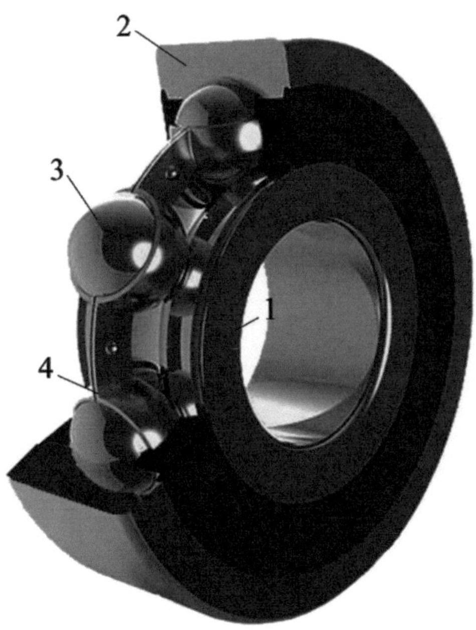

Figure 1.
Components of radial ball bearing.

other. It also maintains proper spacing between the rolling elements. Cages can be made of materials such as steel, brass, or plastic.

Radial ball bearings are commonly used in various applications, including automotive, industrial machinery, and transportation equipment. They offer smooth motion, low friction, and some tolerance for axial misalignment.

Numerous specialized studies have highlighted a series of phenomena that lead to the appearance of vibrations:

- Changes in the position of the rolling elements within the loaded area, leading to modifications in the positions of the rings with influences on elastic deformations in contacts and clearances [1–3];
- Occurrence of deviations in shape and position in the rolling contacts: frontal and radial runout of the raceways, eccentricity, roughness, and appearance of impurities on the raceway with a continuous random distribution over each unit of raceway length [4, 5];
- Rolling elements moving over local defects or damages (pinches, abrasive wear, etc.) [6];
- Uneven movement of rolling elements due to changes in loads, including passing through the unloaded zone [7, 8];
- Frictions and impacts of rolling elements with the cage or the ring [9].

Additionally, determining the correlations between the geometric characteristics of the inner ring raceway deviations in radial ball bearings is essential for ensuring their performance and durability. Geometric deviations, such as ovality, eccentricity, or surface waviness on the raceway, can significantly influence load distribution and the dynamic behavior of the bearing. Thus, due to the trajectory inflections, the path is traversed under shock conditions between the rolling elements and the raceways [10].

This chapter analyzes the behavior, in terms of vibrations, of the forced damped oscillating system (rolling element/raceway) assuming that the outer ring of the bearing is fixed. **Figure 2** depicts a radial bearing loaded with the load P. The rolling elements have a rolling motion without slipping relative to the raceways (inner and outer rings) and have centers at C_0, C_j, C_n loaded with radial forces F_0, F_j, F_n such that $F_0 = max$ and $F_n = 0$.

Due to the contact deformations caused by the radial forces F_j, the distance between the raceways decreases by the amount δ_j, where: $\delta_j = \delta_{ji} + \delta_{je}$.

In the equation provided above, δ_{ji} and δ_{je} represent the displacement of the raceway in the inner and outer rings of the bearing, respectively, with δ_{oi} and δ_{oe} being their maximum values. The force F acting at point M depends on the angle θ. The magnitude of this force decreases, reaching zero at the point where $\theta_n = \pi/2$. R_i and R_e represent the radius of the inner and outer rings of the bearing, respectively. d_o is the diameter of the rolling element, and Δ_r (the distance between the initial center O_1 of the bearing and O_2 the displaced center) represents the eccentricity due to the load, manifested as radial clearance that, under load, causes longitudinal and transverse vibrations in the bearing.

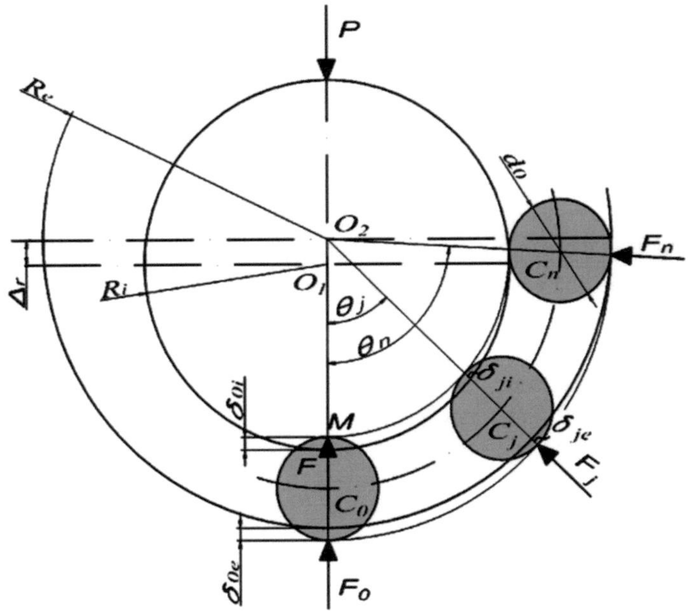

Figure 2.
Radial bearing loaded and the displacement of the raceway in the inner and outer rings of the bearing.

2. Mathematical formulation of dynamic model

The bearing assembly can be seen as two different parts: a mobile component consisting of the balls/rollers, the elastic inner ring, and shaft, and a fixed component, the outer ring. The movements of the bearing elements and shaft lead to a multi-body simulation. The inner ring together with the shaft on the one side and the rolling element on the other side can be seen as a forced damped oscillating system with two degrees of freedom.

Figure 3 presents a schematic representation of radial roller bearing and the simplified mathematical model in which:

- m_1 represents mass of shaft, kg;
- m_2 represents mass of rolling element, kg;
- k_1 represents stiffness coefficient of elastic contact between the inner ring of the bearing and the rolling element, N/m;
- k_2 represents stiffness coefficient of elastic contact between outer ring and rolling element, N/m;
- c_1 represents the damping coefficient due to the conditions between inner ring and rolling element, Ns/m;
- c_2 represents the damping coefficient due to the conditions between outer ring and rolling element, Ns/m.

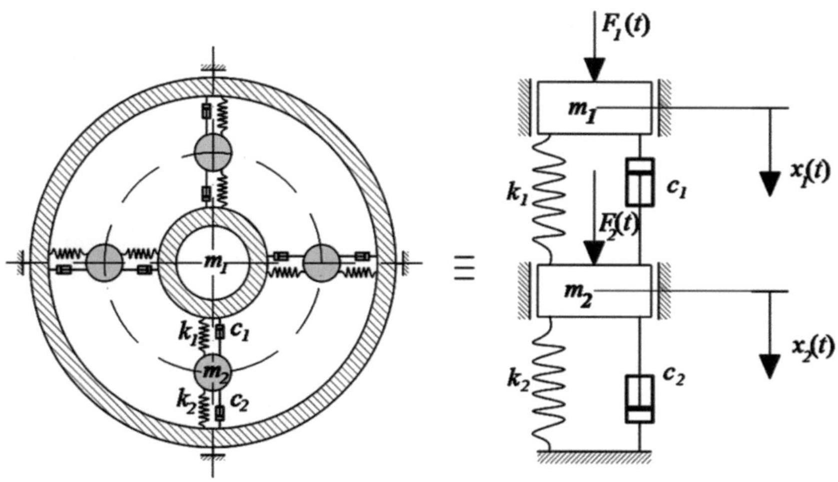

Figure 3.
On the left, schematic representation of radial roller bearing on the right, the simplified mathematical model.

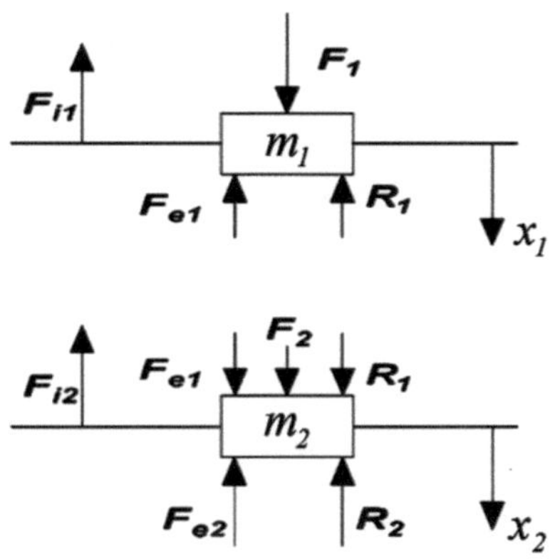

Figure 4.
The two rigids decoupled with forces applied.

For determining dynamic equilibrium equations, the two rigids are decoupled and the forces are applied to them. These forces are represented in **Figure 4**.

The following simplifying notations were used for a better cursive description of the mathematical model:

$$\begin{cases} x(t) = x \\ \dot{x}(t) = \dot{x} \\ \ddot{x}(t) = \ddot{x} \\ F(t) = F \end{cases} \tag{1}$$

The dynamic equilibrium equations are written based on the d'Alembert principle. This principle is applied to each particle in the system, resulting in a set of equations that describe the motion of the entire system:

$$\begin{cases} F_{i1} + R_1 + F_{e1} - F_1 = 0 \\ F_{i2} + R_2 - R_1 + F_{e2} - F_{e1} - F_2 = 0 \end{cases} \tag{2}$$

in which:
$F_{i1,2}$—inertia forces;
$F_{1,2}$—harmonic disturbing forces;
$R_{1,2}$—resisting forces;
$F_{e1,2}$—elastic forces.

$$\begin{cases} F_{i1} = m_1\ddot{x}_1 \\ F_{i2} = m_2\ddot{x}_2 \end{cases};\begin{cases} R_1 = c_1(\dot{x}_1 - \dot{x}_2) \\ R_2 = c_2\dot{x}_2 \end{cases};\begin{cases} F_{e1} = k_1(x_1 - x_2) \\ F_{e2} = k_2x_2 \end{cases};\begin{cases} F_1 = F_{01}e^{i\omega_pt} \\ F_2 = F_{02}e^{i\omega_pt} \end{cases} \tag{3}$$

in which:
$e^{i\omega_pt} = cos\omega_pt + i \cdot sin\,\omega_pt$, ω_p- frequency of the external force.
Substituting Eq. (3) into Eq. (2) gives:

$$\begin{cases} m_1\ddot{x}_1 + c_1(\dot{x}_1 - \dot{x}_2) + k_1(x_1 - x_2) - F_{01}e^{i\omega_pt} = 0 \\ m_2\ddot{x}_2 + c_2\dot{x}_2 - c_1(\dot{x}_1 - \dot{x}_2) + k_2x_2 - k_1(x_1 - x_2) - F_{02}e^{i\omega_pt} = 0 \end{cases} \tag{4}$$

If we make the notation: $e^{i\omega_pt} = e^{rt}$, and we order the terms in Eq. (4) according to de x_1, x_2 we have:

$$\begin{cases} m_1\ddot{x}_1 + c_1\dot{x}_1 - c_1\dot{x}_2 + k_1x_1 - k_1x_2 = F_{01}e^{rt} \\ m_2\ddot{x}_2 - c_1\dot{x}_1 + (c_2 + c_1)\dot{x}_2 - k_1x_1 + (k_2 + k_1)x_2 = F_{02}e^{rt} \end{cases} \tag{5}$$

Writing Eq. (5) in matrix form has the expression:

$$\begin{bmatrix} m_1 & 0 \\ 0 & m_2 \end{bmatrix} \cdot \begin{Bmatrix} \ddot{x}_1 \\ \ddot{x}_2 \end{Bmatrix} + \begin{bmatrix} c_1 & -c_1 \\ -c_1 & c_2 + c_1 \end{bmatrix} \cdot \begin{Bmatrix} \dot{x}_1 \\ \dot{x}_2 \end{Bmatrix} + \begin{bmatrix} k_1 & -k_1 \\ -k_1 & k_2 + k_1 \end{bmatrix} \cdot \begin{Bmatrix} x_1 \\ x_2 \end{Bmatrix} = \begin{Bmatrix} F_{01} \cdot e^{rt} \\ F_{02} \cdot e^{rt} \end{Bmatrix} \tag{6}$$

Reducing Eq. (6) to its simplest form yields:

$$[\mathbf{M}] \cdot \{\ddot{\mathbf{X}}\} + [\mathbf{C}] \cdot \{\dot{\mathbf{X}}\} + [\mathbf{K}] \cdot \{\mathbf{X}\} = \{\mathbf{F}\} \tag{7}$$

In the above relation, the vectors of accelerations, velocities, and displacements have the form:

$$\{\mathbf{X}\} = \begin{Bmatrix} x_1 \\ x_2 \end{Bmatrix}; \quad \{\dot{\mathbf{X}}\} = \begin{Bmatrix} \dot{x}_1 \\ \dot{x}_2 \end{Bmatrix}; \quad \{\ddot{\mathbf{X}}\} = \begin{Bmatrix} \ddot{x}_1 \\ \ddot{x}_2 \end{Bmatrix} \tag{8}$$

If we denote the coefficients of the stiffness, mass, and damping matrices as a_{ij}, their form becomes:

$$[\mathbf{M}] = \begin{bmatrix} m_{11} & 0 \\ 0 & m_{22} \end{bmatrix}; [\mathbf{C}] = \begin{bmatrix} c_{11} & c_{12} \\ c_{21} & c_{22} \end{bmatrix}; [\mathbf{K}] = \begin{bmatrix} k_{11} & k_{12} \\ k_{21} & k_{22} \end{bmatrix}; \{\mathbf{F}\} = \begin{Bmatrix} F_1 \\ F_2 \end{Bmatrix} \tag{9}$$

It can be observed that, in the analyzed case, there is a particularity, namely: $c_{12} = c_{21}$ și $k_{12} = k_{21}$.
With the above notations, Eq. (6) becomes:

$$\begin{bmatrix} m_{11} & 0 \\ 0 & m_{22} \end{bmatrix} \cdot \begin{Bmatrix} \ddot{x}_1 \\ \ddot{x}_2 \end{Bmatrix} + \begin{bmatrix} c_{11} & c_{12} \\ c_{21} & c_{22} \end{bmatrix} \cdot \begin{Bmatrix} \dot{x}_1 \\ \dot{x}_2 \end{Bmatrix} + \begin{bmatrix} k_{11} & k_{12} \\ k_{21} & k_{22} \end{bmatrix} \cdot \begin{Bmatrix} x_1 \\ x_2 \end{Bmatrix} = \begin{Bmatrix} F_1 \\ F_2 \end{Bmatrix} \tag{10}$$

Expanding the above system yields:

$$\begin{cases} m_{11}\ddot{x}_1 + c_{11}\dot{x}_1 + c_{12}\dot{x}_2 + k_{11}x_1 + k_{12}x_2 = F_{01}e^{rt} \\ m_{22}\ddot{x}_2 + c_{21}\dot{x}_1 + c_{22}\dot{x}_2 + k_{21}x_1 + k_{22}x_2 = F_{02}e^{rt} \end{cases} \tag{11}$$

The particular solutions of Eq. (11) with first- and second-order derivatives have the form:

$$\begin{cases} x_1 = A_1e^{st} + B_1e^{rt} \\ \dot{x}_1 = A_1se^{st} + B_1re^{rt} \\ \ddot{x}_1 = A_1s^2e^{st} + B_1r^2e^{rt} \end{cases} ; \quad \begin{cases} x_2 = A_2e^{st} + B_2e^{rt} \\ \dot{x}_2 = A_2se^{st} + B_2re^{rt} \\ \ddot{x}_2 = A_2s^2e^{st} + B_2r^2e^{rt} \end{cases} \tag{12}$$

in which:

- Coefficients $A_{1,2}$ and $B_{1,2}$ represent coefficients characterizing the amplitude in free and forced oscillation, respectively;

- $e^{i\omega_n t} = cos\omega_n t + i \cdot sin\,\omega_n t$, ω_n- natural frequency of the system.

For simplicity we make the notation: $e^{i\omega_n t} = e^{st}$.

Eq. (12) is substituted into system Eq. (11), then grouped by e^{st} and e^{rt}. Afterward grouping by the coefficients $A_{1,2}$ and $B_{1,2}$ yields:

$$\begin{cases} \left[(m_{11}s^2 + c_{11}s + k_{11})A_1 + (c_{12}s + k_{12})A_2\right]e^{st} + \\ + \left[(m_{11}r^2 + c_{11}r + k_{11})B_1 + (c_{12}r + k_{12})B_2\right]e^{rt} = F_{01}e^{rt} \\ \left[(m_{22}s^2 + c_{22}s + k_{22})A_2 + (c_{21}s + k_{21})A_1\right]e^{st} + \\ + \left[(m_{22}r^2 + c_{22}r + k_{22})B_2 + (c_{21}r + k_{21})B_1\right]e^{rt} = F_{02}e^{rt} \end{cases} \tag{13}$$

Identifying term by term in Eq. (13) and grouping by $e^{st}$and e^{rt} (yields)

$$\begin{cases} \left[(m_{11}s^2 + c_{11}s + k_{11})A_1 + (c_{12}s + k_{12})A_2\right]e^{st} = 0 \\ \left[(m_{22}s^2 + c_{22}s + k_{22})A_2 + (c_{21}s + k_{21})A_1\right]e^{st} = 0 \\ \left[(m_{11}r^2 + c_{11}r + k_{11})B_1 + (c_{12}r + k_{12})B_2\right]e^{rt} = F_{01}e^{rt} \\ \left[(m_{22}r^2 + c_{22}r + k_{22})B_2 + (c_{21}r + k_{21})B_1\right]e^{rt} = F_{02}e^{rt} \end{cases} \tag{14}$$

The system Eq. (14) contains two subsystems that can be solved independently.

2.1 Solving subsystem I

$$\begin{cases} \left[(m_{11}s^2 + c_{11}s + k_{11})A_1 + (c_{12}s + k_{12})A_2\right]e^{st} = 0 \\ \left[(m_{22}s^2 + c_{22}s + k_{22})A_2 + (c_{21}s + k_{21})A_1\right]e^{st} = 0 \end{cases} \tag{15}$$

For $e^{st} \neq 0$, yields:

$$\begin{cases} (m_{11}s^2 + c_{11}s + k_{11})A_1 + (c_{12}s + k_{12})A_2 = 0 \\ (c_{21}s + k_{21})A_1 + (m_{22}s^2 + c_{22}s + k_{22})A_2 = 0 \end{cases} \tag{16}$$

Written in matrix form, system Eq. (16) becomes:

$$[\mathbf{A_s}] \cdot \begin{Bmatrix} A_1 \\ A_2 \end{Bmatrix} = \begin{Bmatrix} 0 \\ 0 \end{Bmatrix} \tag{17}$$

in which:

$$[\mathbf{A_s}] = \begin{bmatrix} m_{11}s^2 + c_{11}s + k_{11} & c_{12}s + k_{12} \\ c_{21}s + k_{21} & m_{22}s^2 + c_{22}s + k_{22} \end{bmatrix} = \begin{bmatrix} A_{s_{11}} & A_{s_{12}} \\ A_{s_{21}} & A_{s_{22}} \end{bmatrix} \tag{18}$$

The Eq. (17) becomes:

$$\begin{bmatrix} A_{s_{11}} & A_{s_{12}} \\ A_{s_{21}} & A_{s_{22}} \end{bmatrix} \cdot \begin{Bmatrix} A_1 \\ A_2 \end{Bmatrix} = \begin{Bmatrix} 0 \\ 0 \end{Bmatrix} \tag{19}$$

$$\begin{Bmatrix} A_1 \\ A_2 \end{Bmatrix} = [\mathbf{A_s}]^{-1} \cdot \begin{Bmatrix} 0 \\ 0 \end{Bmatrix} = \frac{1}{\det[\mathbf{A_s}]} \cdot [\mathbf{A_s}]^* \cdot \begin{Bmatrix} 0 \\ 0 \end{Bmatrix} \tag{20}$$

In Eq. (20) for $\det[\mathbf{A_s}] \neq 0$trivialsolutions are obtained $(A_{1,2} = 0)$. Eq. (20) holds true until $\det[\mathbf{A_s}] = 0$. If the condition $\det[\mathbf{A_s}] = 0$ is imposed, the characteristic polynomial (Eq. (22)) from which the solutions $s_{1\ldots4}$ containing the eigenfrequencies result:

$$\det[\mathbf{A_s}] = 0 \rightarrow \begin{vmatrix} m_{11}s^2 + c_{11}s + k_{11} & c_{12}s + k_{12} \\ c_{21}s + k_{21} & m_{22}s^2 + c_{22}s + k_{22} \end{vmatrix} = 0 \tag{21}$$

$$\begin{gathered}(m_{11}m_{22}s^4) + (m_{11}c_{22} + m_{22}c_{11})s^3 + (m_{11}k_{22} + m_{22}k_{11} - (c_{12}c_{21}) + (c_{11}c_{22}))s^2 + \\ (c_{11}k_{22} + c_{22}k_{11} - c_{12}k_{21} - c_{21}k_{12})s + (k_{11}k_{22} - k_{12}k_{21}) = 0\end{gathered} \tag{22}$$

For an oscillating system with two degrees of freedom, such as the one analyzed, the solutions of the characteristic polynomial will be four. These roots are generally complex and appear in conjugate pairs due to the nature of the damping. The form of these roots has the expression:

$$\begin{aligned} s_{12} &= -\alpha_1 \pm \omega_{d1} \\ s_{34} &= -\alpha_2 \pm \omega_{d2} \end{aligned} \tag{23}$$

in which:

α_{12} represents the damping coefficients and ω_{d12} represents damped oscillations.

Based on the damping coefficients and damped oscillations, the natural frequencies ω_{n12} result:

$$\begin{aligned} \omega_{n1} &= \sqrt{\alpha_1{}^2 + \omega_{d1}{}^2} \\ \omega_{n2} &= \sqrt{\alpha_2{}^2 + \omega_{d2}{}^2} \end{aligned} \tag{24}$$

The solutions for the coefficients $A_{1,2}$ are obtained based on Cramer's rule from Eq. (19):

$$\begin{cases} A_1 = A_{s_{22}} - A_{s_{12}} \\ A_2 = A_{s_{11}} - A_{s_{21}} \end{cases} \tag{25}$$

2.2 Solving subsystem II

$$\begin{cases} \left(m_{11}r^2 + c_{11}r + k_{11}\right)B_1 + (c_{12}r + k_{12})B_2 = F_{01} \\ (c_{21}r + k_{21})B_1 + \left(m_{22}r^2 + c_{22}r + k_{22}\right)B_2 = F_{02} \end{cases} \tag{26}$$

Written in matrix form, system Eq. (26) becomes:

$$[\mathbf{B_r}] \cdot \begin{Bmatrix} B_1 \\ B_2 \end{Bmatrix} = \begin{Bmatrix} F_{01} \\ F_{02} \end{Bmatrix} \tag{27}$$

in which:

$$[\mathbf{B_r}] = \begin{bmatrix} m_{11}r^2 + c_{11}r + k_{11} & c_{12}r + k_{12} \\ c_{21}r + k_{21} & m_{22}r^2 + c_{22}r + k_{22} \end{bmatrix} = \begin{bmatrix} B_{r_{11}} & B_{r_{12}} \\ B_{r_{21}} & B_{r_{22}} \end{bmatrix} \tag{28}$$

Eq. (27) becomes:

$$\begin{bmatrix} B_{r_{11}} & B_{r_{12}} \\ B_{r_{21}} & B_{r_{22}} \end{bmatrix} \cdot \begin{Bmatrix} B_1 \\ B_2 \end{Bmatrix} = \begin{Bmatrix} F_{01} \\ F_{02} \end{Bmatrix} \tag{29}$$

The solutions for the coefficients $B_{1,2}$ are obtained based on Cramer's rule with the condition: $\det[\mathbf{B_r}] \neq 0$.

$$B_1 = \frac{F_{01} \cdot B_{r_{22}} - F_{02} \cdot B_{r_{12}}}{B_{r_{11}} \cdot B_{r_{22}} - B_{r_{12}} \cdot B_{r_{21}}} \tag{30}$$

$$B_2 = \frac{F_{02} \cdot B_{r_{11}} - F_{01} \cdot B_{r_{21}}}{B_{r_{11}} \cdot B_{r_{22}} - B_{r_{12}} \cdot B_{r_{21}}} \tag{31}$$

Based on the particular solutions, with the initial condition: $t = 0 \rightarrow x_1(0) = x_{01}$, $x_2(0) = x_{02}$, $\dot{x_1}(0) = v_{01}$, $\dot{x_2}(0) = v_{02}$, the displacements and velocities of the two masses of the system will result.

$$\begin{cases} x_{01} = A_1 + B_1 \\ x_{02} = A_2 + B_2 \end{cases} \tag{32}$$

$$\begin{cases} v_{01} = A_1 s + B_1 r \\ v_{02} = A_2 s + B_2 r \end{cases} \tag{33}$$

3. Results of representative example

It further integrated this mathematical method into MATLAB for solving different combinations of input parameters. By integrating this method into MATLAB, we can leverage the computational capabilities of the software to efficiently solve such a system of differential equations and analyze the behavior of dynamic systems. The algorithm consists of defining the differential equation, finding the general solution, determining integration constants using initial conditions, and obtaining the specific solution.

For the same input parameters and the amplitude of the external forces with different values of the excitation frequencies, the graphs below were obtained (**Figures 5–8**).

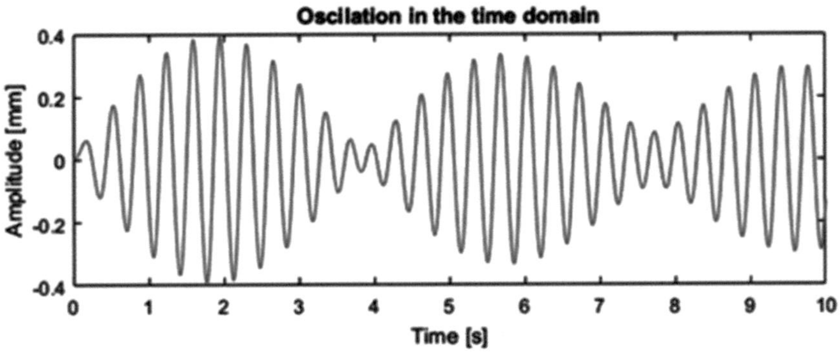

Figure 5.
System response for excitation frequencies, $\omega_p = 18.5\ rad/s$.

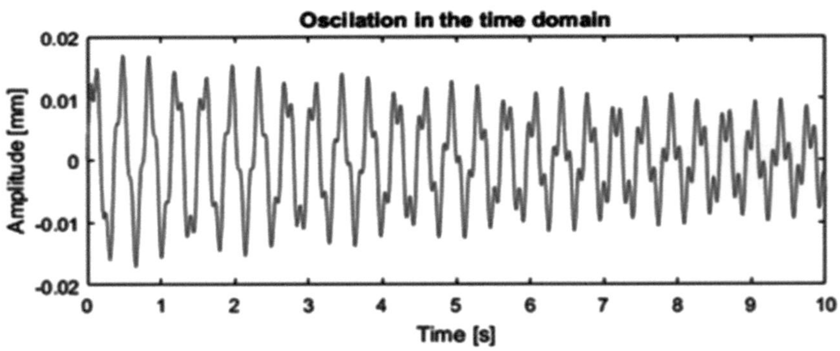

Figure 6.
System response for excitation frequencies, $\omega_p = 55\ rad/s$.

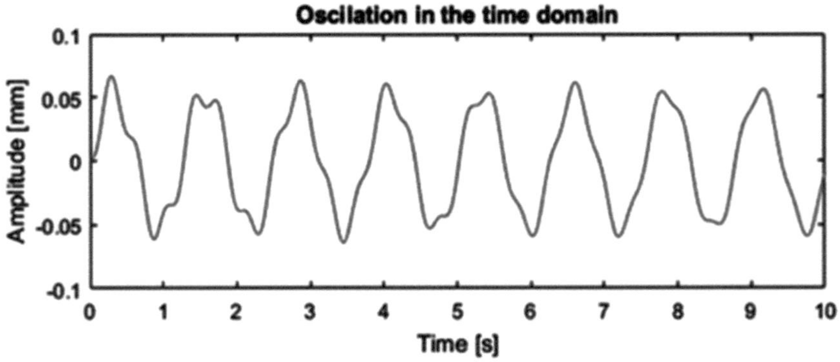

Figure 7.
System response for excitation frequencies, $\omega_p = 5\ rad/s$.

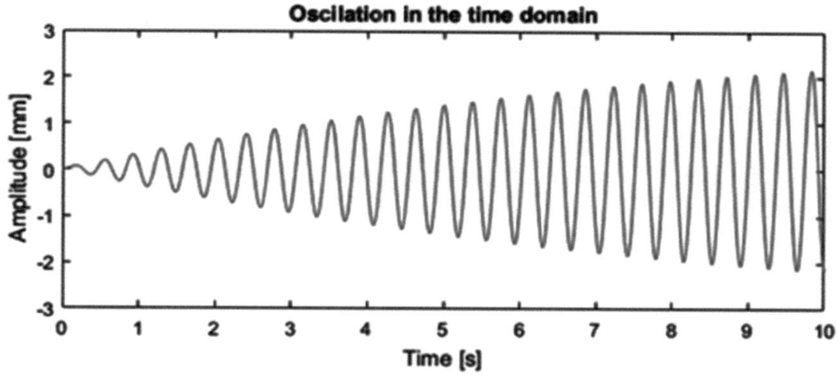

Figure 8.
System response for excitation frequencies, $\omega_p = 17\ rad/s$.

The behavior of a mechanical system, as described in the previous algorithm, is significantly influenced by its main parameters: mass, damping coefficients, stiffness constants, and external forces. Each of these parameters has a crucial role in determining the response of the system to excitations. Here is how these parameters influence system behavior:

3.1 Masses in the system

Mass confers inertia on a body, i.e. its resistance to changes in motion. Higher mass leads to lower natural frequencies because the natural frequency is proportional to k. Dynamically, a larger mass tends to reduce the vibration amplitude for a given excitation, due to the increased inertia.

3.2 Damping coefficients

Damping coefficients represent the forces that resist motion, dissipating the system's energy, usually in the form of heat. A higher damping coefficient reduces the duration of the oscillations over time, making the system stabilize faster. Damping also affects the amplitude of the steady-state response and at resonant frequencies, where high damping can prevent excessive amplitude growth.

3.3 Stiffness constants

Stiffness constants measure the resistance to elastic deformation in the system. The natural frequencies are directly proportional to the square root of the stiffness constant. Therefore, increasing the stiffness will increase the natural frequencies, making the system vibrate faster. Stiffness also influences the overall response of the system to excitations. A higher stiffness leads to a smaller increase in displacement for a given force, which means that the system is more "stiff."

3.4 External forces and their frequency

If the excitation frequency approaches one of the system's natural frequencies, the resonance phenomenon may occur, leading to large vibration amplitudes. The amplitude of the response is inversely proportional to the damping and can become dangerously large under resonant conditions. On the other hand, the magnitude of the external force directly influences the amplitude of the system response. Larger external forces induce larger amplitude responses.

The flowchart below provides a clear visual representation of the algorithm's steps and decision points.

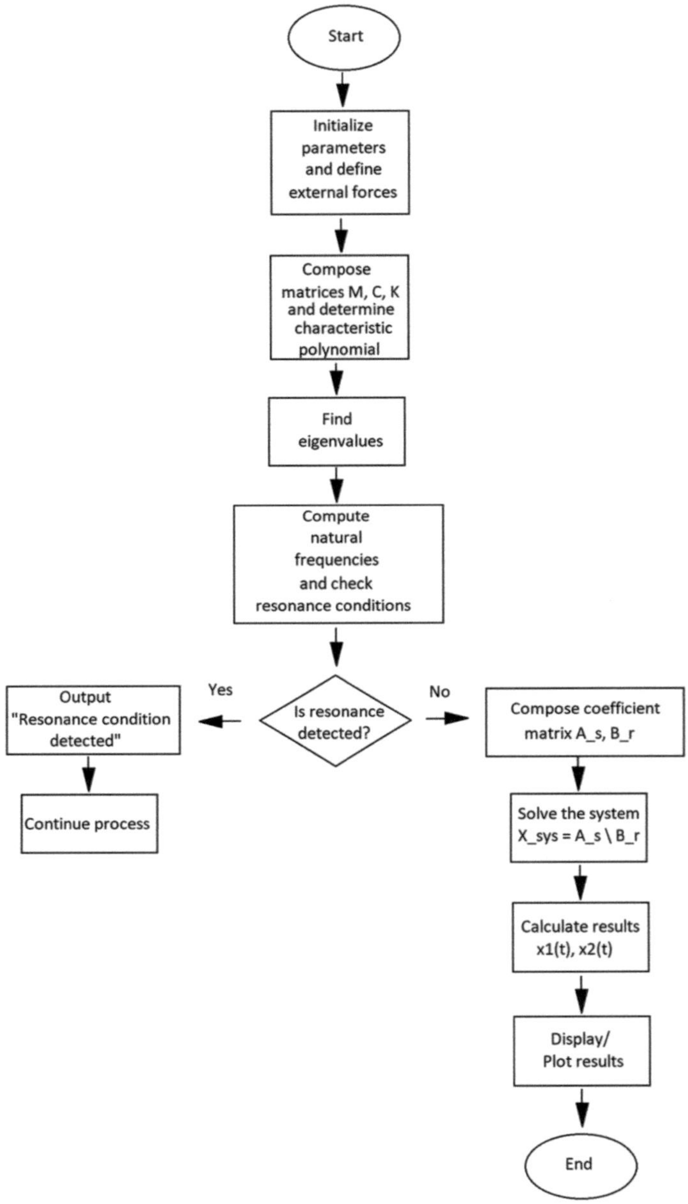

Below, complete script file is included for random values of parameters.

```
% Define input parameters
t = 0:0.01:10; % Time vector
```

```
m1 = 15; % Mass of the first body
m2 = 0.065; % Mass of the second body
c1 = 20; % Damping coefficient of the first body
c2 = 15; % Damping coefficient of the second body
k1 = 20^2; % Stiffness constant of the first spring
k2 = 15^2; % Stiffness constant of the second spring
% Define external forces
F01 = 1000; % Amplitude of external force on the first mass
F02 = 2000; % Amplitude of external force on the second mass
r = 20; % Frequency of external forces (rad/s)
% Composing mass, damping, and stiffness matrices
M = [m1 0; 0 m2];
C = [c1 + c2 −c2; −c2 c2];
K = [k1 + k2 −k2; −k2 k2];
% Determining characteristic polynomial
syms s
P_s = det(M*s^2 + C*s + K);
% Solving for eigenvalues (natural frequencies)
eigenvalues = double(solve(P_s == 0, s));
wn_1 = sqrt(real(eigenvalues(1))^2 + imag(eigenvalues(1))^2);
wn_2 = sqrt(real(eigenvalues(2))^2 + imag(eigenvalues(2))^2);
% Check for resonance conditions
if abs(r - wn_1) < 1e-3 || abs(r - wn_2) < 1e-3
disp('Resonance condition detected.');
end
% Composing Coefficient Matrix for steady-state response
omega = r; % Assuming the system is excited at frequency r
A_s = M*(1i*omega)^2 + C*(1i*omega) + K;
% Vector for the external forces
B_r = [F01; F02] * exp.(1i*omega*t); % This creates a 2xN matrix, where N is the
number of time points
% Solve the system for each time point
X_sys = A_s\B_r; % X_sys will be a 2xN matrix
% Results
% Extract the time-domain responses for both masses
x1 = real(X_sys(1,:));
x2 = real(X_sys(2,:));
% Plotting the results
figure;
subplot(2, 1, 1);
plot(t, x1);
title('Response of the first mass');
xlabel('Time (s)');
ylabel('Displacement (m)');
subplot(2, 1, 2);
plot(t, x2);
title('Response of the second mass');
xlabel('Time (s)');
ylabel('Displacement (m)');
```

4. Conclusion

The characteristic polynomial, derived from the determinant of the system matrix containing its eigenvalues, plays an essential role in this analysis. It not only describes the system's eigenvalues but also the relationship between damping and its dynamic behavior. Thus, similar to how polynomials reveal the roots and key points of mathematical functions, the characteristic polynomial provides crucial information about the dynamics of the damped mechanical vibration system. In conclusion, polynomials form a conceptual bridge between abstract mathematics and the analysis of mechanical vibrations, demonstrating once again the applicability of mathematical models in understanding and modeling the physical world. The proposed mathematical model is versatile, offering a broad application scope to any analogous oscillatory system. The behavior of a mechanical system is complex and depends on the interaction between system parameters. By adjusting these, natural frequencies, damping response, and stiffness can be controlled, and dangerous resonant conditions can be avoided. In systems design and analysis, it is essential to understand these influences to ensure optimal system stability and performance.

Conflict of interest

The author declares no conflict of interest.

Author details

Lucian Milica
Faculty of Engineering, "Dunarea de Jos" University of Galati, Galati, Romania

*Address all correspondence to: milica.lucian@ugal.ro

References

[1] Harris TA, Kotzalas MN. Essential Concepts of Bearing Technology. 5th ed. Boca Raton, FL, USA: CRC Press; 2006. DOI: 10.1201/9781420006599

[2] Cao H, Niu L, Xi S, Chen X. Mechanical model development of rolling bearing-rotor systems: A review. Mechanical Systems and Signal Processing. 2018;**102**:37-58. DOI: 10.1016/j.ymssp.2017.09.023

[3] Sharma A, Upadhyay N, Kankar PK, Amarnath M. Nonlinear dynamic investigations on rolling element bearings: A review. Advances in Mechanical Engineering. 2018;**10**(3): 1-15. DOI: 10.1177/1687814018764148

[4] Yakout M, Nassef MGA, Backar S. Effect of clearances in rolling element bearings on their dynamic performance, quality and operating life. Journal of Mechanical Science and Technology. 2019;**33**:2037-2042. DOI: 10.1007/s12206-019-0406-y

[5] Zhang SY, Spiryagin M, Ding HH, Wu Q, Guo J, Liu QY, et al. Rail rolling contact fatigue formation and evolution with surface defects. International Journal of Fatigue. 2022;**158**:130-138. DOI: 10.1016/j.ijfatigue.2022.106762

[6] Xiong Q, Li Y, Wang Z, Xu Y, Zhu Y, Gu F, et al. Dynamic models for local faults on rolling element bearings: A review. In: Proceedings of IncoME-V & CEPE Net-2020. Cham: Springer; 2021. pp. 217-227. DOI: 10.1007/978-3-030-75793-9_22

[7] Tu W, Yang J, Luo Y. Contact characteristic and vibration mechanism of rolling element bearing in the process of fault evolution. Proceedings of the Institution of Mechanical Engineers, Part K: Journal of Multi-Body Dynamics. 2021;**235**:19-36. DOI: 10.1177/1464419320985707

[8] Chen J, Liu J, Shao Y, Luo C. Vibration modeling of lubricated rolling element bearing considering skidding in loaded zone. Journal of Failure Analysis and Prevention. 2014;**14**:809-817. DOI: 10.1007/s11668-014-9895-2

[9] Zhao D, Hong J, Yan K, Zhao Q, Fang B. Dynamic interaction between the rolling element and cage of rolling bearing considering cage flexibility and clearance. Mechanism and Machine Theory. 2022;**174**:298-308. DOI: 10.1016/j.mechmachtheory.2022.104905

[10] Halminen O, Aceituno J, Escalona JL, Sopanen J, Mikkola A. A touchdown bearing with surface waviness: Friction loss analysis. Mechanism and Machine Theory. 2017;**110**:73-84. DOI: 10.1016/j.mechmachtheory.2017.01.002